THE CRYOSPHERE

THE CRYOSPHERE

Shawn J. Marshall

PRINCETON UNIVERSITY PRESS   *Princeton & Oxford*

Published by Princeton University Press
41 William Street, Princeton, New Jersey 08540
In the United Kingdom: Princeton University Press
6 Oxford Street, Woodstock, Oxfordshire OX20 1TW
press.princeton.edu

ISBN 978-0-691-14525-9
ISBN (pbk.) 978-0-691-14526-6

Library of Congress Cataloging-in-Publication Data

Marshall, Shawn.
The cryosphere / Shawn J. Marshall.
p. cm. — (Princeton primers in climate)
Includes bibliographical references and index.
ISBN 978-0-691-14525-9 (hardcover) — ISBN 978-0-691-14526-6 (pbk.)
1. Cryosphere. 2. Climatic changes. I. Title.
QC880.4.C79M73 2011
551.31—dc23
2011017995

British Library Cataloging-in-Publication Data is available
This book has been composed in Avenir and Minion Pro

10 9 8 7 6 5 4 3 2 1

Contents

Preface

ALL CANADIANS COME INTO A WORLD THAT IS SHAPED by snow and ice, and I am exceptionally privileged in this regard. I was born in the heart of winter in Matheson, Ontario, a small mining town in the northern part of the province where one can count on snow cover from Halloween to Mother's Day. For half a year the snow is part of your fabric, and for the other half of the year one can bike, run, and paddle the landforms and lakes formed by the great ice sheets that carved the landscape. I took up Nordic ski racing, where I became a student of the subtle influences of snow temperature, moisture, and texture on grip and glide. I have enjoyed the company of too many friends to name (and I include my graduate students in this) on ski trips and in glacier field work. Faron Anslow, Joe Shea, Tara Moran, Kate Sinclair, Gwenn Flowers, Dave Hildes, and Phil Hammer stand out within this group. Thanks also to several pagophilic friends and colleagues that helped with this text, including Jacqueline Dumas, Eric White, Cecilia Bitz, Camille Li, and Tom Lambert.

Professionally, I am indebted to many colleagues for exposing me to the scientific wonders of snow and ice. As an undergraduate student at the University of Toronto, I had a fortunate confluence with Dick Peltier, who loaned me a copy of Imbrie and Imbrie's *Ice Ages: Solving the Mystery*. This exposed me to the incredible rhythms and unresolved mysteries of Ice Age cycles: Earth's definitive testimony to the importance of the cryosphere in climate dynamics. I went on to study glaciology at the University of British Columbia, where I enjoyed graduate courses from J. Ross McKay, Dave McClung, and Garry Clarke in permafrost, avalanche processes, and glacier dynamics. Garry, my Ph.D. mentor, brought me to Trapridge Glacier in the Yukon: a magical setting, one of the best-studied glaciers in the world, and the incubator for many of the ideas, instruments, and research methods that underlie current understanding of glacier dynamics. Helgi Björnsson entrained me in studies of ice caps in Iceland, where fire and ice clash with spectacular results. Martin Sharp invited me to field studies on Ellesmere Island in the Canadian high Arctic, where the spring landscape is a brilliant white as far as the eye can see.

This text is intended as a brief introduction to the topic, but I hope that these pages capture the essential physics and character of the cryosphere and inspire others to further exploration of the cryosphere's role in Earth's climate. Although I have seen snow and ice in many guises, I am still a student of cryosphere dynamics. I have striven to provide a balanced perspective on all aspects of the global cryosphere, but I suspect that the depth of my understanding of certain issues, and lack thereof, shines through in places. The suggestions for further reading provide greater detail on all topics and will help to fill gaps in my coverage. Sincere thanks to

Bob Bindschadler, Koni Steffen, and two anonymous reviewers for their suggestions, which have improved this text. Alison Kalett at Princeton University Press has been a delight to work with, and I thank Alison and the production staff at the Press for their enthusiasm, support, and flexibility.

Lawren S. Harris: *Ellesmere Island* (1930). (The McMichael Canadian Art Collection, Kleinberg, Ontario, Canada. Gift of Mrs. Chester Harris.)

THE CRYOSPHERE

1 INTRODUCTION TO THE CRYOSPHERE

In this place, nostalgia
roams, patient as slow
hands on skin, transparent
as melt-water. Nights are light
and long. Shadows settle
on the shoulders of air.
Time steps out of line
here, stops to thaw
the frozen hearts of icebergs.
Sleep isn't always easy in this place
where the sun stays up all night
and silence has a voice.
—Claire Beynon, "At Home in Antarctica"

EARTH SURFACE TEMPERATURES ARE CLOSE TO THE triple point of water, 273.16 K, the temperature at which water vapor, liquid water, and ice coexist in thermodynamic equilibrium. Indeed, water is the only substance on Earth that is found naturally in all three of its phases. Approximately 35% of the world experiences temperatures below the triple point at some time in the year, including about half of Earth's land mass, promoting frozen water at Earth's surface. The global cryosphere encompasses all aspects of this frozen realm, including glaciers and ice sheets, sea ice, lake and river ice, permafrost, seasonal snow, and ice crystals in the atmosphere.

Because temperatures oscillate about the freezing point over much of the Earth, the cryosphere is particularly sensitive to changes in global mean temperature. In a tight coupling that represents one of the strongest feedback systems on the planet, global climate is also directly affected by the state of the cryosphere. Earth temperatures are primarily governed by the net radiation that is available from the Sun. Because solar variability is modest on annual to million-year timescales (less than 1% of the solar constant), the single most dynamic control of net radiation is the global albedo—the planetary reflectivity—which is heavily influenced by the areal extent of snow and ice covering the Earth. The simple but illuminating global climate models of Mikhail Budyko and William Sellers explored this feedback in the late 1960s, demonstrating the delicate balance between Earth's climate and cryosphere.

GEOGRAPHY OF EARTH'S SNOW AND ICE

Perennial ice covers 10.8% of Earth's land surface (table 1.1 and figure 1.1), with most of this ice stored in the great polar ice sheets in Greenland and Antarctica. Smaller glaciers and icefields are numerous—the global population is estimated at more than 200,000—but these ice masses cover a relatively small area of the landscape. An additional 15.4% of Earth's land surface is covered by

Table 1.1
Area and Volume of the Global Cryosphere

	Perennial Ice		
Location	Area $(10^6 \, km^2)$	Volume $(10^6 \, km^3)$	Sea Level Equivalent $(m)^a$
Greenland ice sheet	1.7	2.9	7.1
Antarctic ice sheet[b]	13.3	25.4	56.2
Mountain glaciers[c]	1.1	0.22–0.38	0.56–0.97
Permafrost	22.8	0.01–0.04	0.03–0.10

	Snow and Sea Ice					
Type and	Area $(10^6 \, km^2)$			Extent[d] $(10^6 \, km^2)$		
Hemisphere	Minimum	Maximum	Mean	Minimum	Maximum	Mean
Sea ice						
Northern	4.8	13.6	9.8	6.6	15.5	11.8
Southern	1.9	14.5	8.7	3.0	18.8	12.0
Snow						
Northern				3.1	46.7	24.9
Southern				13.9	15.1	14.3

Note: Sea ice data is from January 1979 to March 2011, and Northern Hemisphere snow data is from November 1966 to March 2011.

[a]Sea level equivalent is the increase in global average eustatic sea level that would occur if all of this ice was transferred to the oceans.

[b]Includes marine-based (floating) ice shelves, but excludes peripheral icefields.

[c]Mountain glaciers and polar ice caps, excluding the Greenland and Antarctic ice sheets but including dynamically independent icefields peripheral to each ice sheet (see chapter 6).

[d]*Extent* is the total area of the region experiencing snow or ice cover. For sea ice, this includes all satellite pixels (with a resolution of ca. 625 km^2) with an ice concentration of at least 15%; the remainder of the area in a pixel can be land or open water. Snow extent is defined by the edge of the region with near-continuous snow cover; there can be snow-free patches within this region.

permafrost: frozen ground that ranges from a few meters to hundreds of meters deep.

In contrast to this permanent ice, seasonal snow and ice fluctuate dramatically. Snow cover is the most variable element of the cryosphere. From 1966 to 2011, the Northern Hemisphere winter snow cover reached an average maximum extent of 46.7×10^6 km^2: almost half of the Northern Hemisphere land mass (figure 1.3). There is almost complete loss of this snow each summer, with permanent snow cover limited to the interior of Greenland and the accumulation areas of other high-altitude and polar ice caps.

Because the Southern Hemisphere continents are situated at lower latitudes (excepting Antarctica), southern snow cover is less extensive. It is also less studied, with satellite composite images of total snow-covered area available only since 2000. The South American Andes, high elevations of southeastern Australia, much of New Zealand, and the islands off of Antarctica all experience seasonal snow, as do the high peaks in tropical East Africa. Based on the July 0°C isotherm, the total area of this maximum snow cover is estimated to be 1.2×10^6 km^2, with most of this snow residing in the Patagonian icefields

Figure 1.1. Global (a) Arctic and (b) Antarctic sea ice cover, February 1, 2011. Also shown are the Greenland and Antarctic Ice Sheets. Data provided by the National Center for Environmental Prediction/NOAA and the U.S. National Snow and Ice Data Center. Images from The Cryosphere Today, University of Illinois at Urbana-Champaign (http://arctic.atmos.uiuc.edu/cryosphere/).

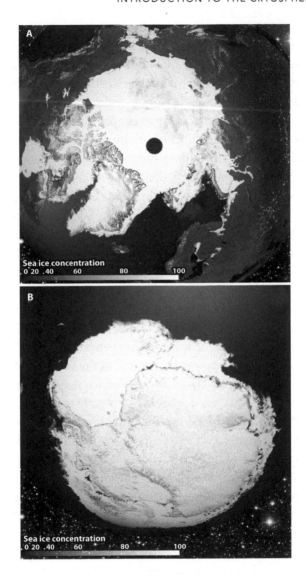

of South America. Combined with the permanent blanket of snow over Antarctica, this gives a peak Southern Hemisphere terrestrial snow cover of 15.1×10^6 km^2, approximately one-third that of the Northern Hemisphere.

There is less of a seasonal cycle for the Southern Hemisphere snowpack, as most of Antarctica is too cold to experience summer melting. Snows are perennial across the frozen continent, with melting confined to the coastal periphery. As a tangential but delightful consequence of this, earthshine is exceptionally bright in December and January, when the Sun is sojourning in the Southern Hemisphere and reflected sunlight from Antarctica adds to the solar illumination of the Moon. In a sense, everyone in the world can see the Antarctic snows in the lunar orb.

The white blanket that spreads over the land surface each winter has a direct parallel in the high-latitude oceans, where sea ice forms a thin veneer that effectively transforms water to land for much of the year. Figure 1.2 illustrates a "field" of snow-covered ice floes aligned by the wind during sea-ice breakup in early summer (June 2005). Sea ice is made up of a combination of first-year and multiyear ice. First-year ice forms anew from in situ freezing of seawater each year. Multiyear ice has survived at least one summer melt season, persevering through two main mechanisms: (i) some ice remains at high latitudes as a result of being landfast, stuck within a channel or bay, or cycled within ocean gyres that trap rather than export the ice; (ii) ice floes ridge or pile up in areas of convergence, producing thick, resilient ice.

Figure 1.2. Snow-covered sea ice floes and melt ponds during spring breakup, Button Bay (Hudson Bay), Manitoba, Canada. The ice floes are aligned by the wind. Scientific instrumentation (a ground based microwave scatterometer) is visible in the center of the picture. (Photograph by John Yackel.)

These mechanisms often operate in concert and are more prevalent in the Arctic than the Antarctic, resulting in a thicker ice cover and more multiyear ice in the north.

Relative to the continents, seasonal cycles of ice in the oceans are more hemispherically symmetric (table 1.1), although there are interesting north–south contrasts. Passive microwave remote sensing for the period 1979–2011 indicates an average minimum Northern Hemisphere ice area of 4.8×10^6 km^2, typically reached in September. Maximum ice cover is usually attained in late winter, with an average March ice-covered area of 13.6×10^6 km^2. Sea ice in the Southern Ocean has a larger seasonal cycle, with relatively little multiyear ice. Annual mean sea-ice cover in the south is 8.7×10^6 km^2, varying from 1.9×10^6 km^2 (February) to 14.5×10^6 km^2 (September).

Combining the hemispheres, global sea-ice area is relatively constant, varying from 15.4×10^6 to 20.8×10^6 km^2, with a minimum in February and a peak in November. Global ice extent—the area of the oceans containing sea ice, as demarcated by the ice edge—varies from 18.4×10^6 to 27.3×10^6 km^2. Mean annual global ice area and extent are 18.5×10^6 and 23.9×10^6 km^2.

Combining the snow and sea-ice cover, the seasonal cryosphere blankets 59×10^6 and 30×10^6 km^2 in the Northern Hemisphere and Southern Hemisphere, respectively. Figure 1.3 illustrates the geographic distribution. Additional elements of the cryosphere include seasonally frozen ground and freshwater (river and lake) ice.

This snow and ice cover influences the surface albedo and energy budget of the planet fluxes of heat and moisture between the atmosphere and surface and the patterns of circulation in the ocean and atmosphere. Each element of the global cryosphere interacts with and affects weather, climate, and society, and each is highly sensitive to global climate change.

This book explores the physics and characteristics of the global cryosphere, with an emphasis on cryosphere–climate interactions. Chapter 2 presents an overview of the structure of snow and ice in its various manifestations on Earth, including the material properties that

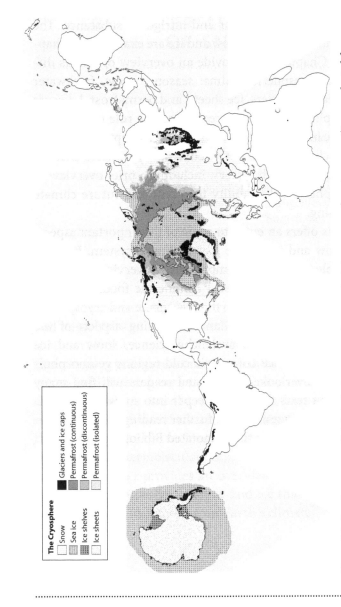

Figure 1.3. Areal coverage of global glaciers, permafrost, and winter snow and sea ice cover. (Image adapted from the UNEP/GRID-Arendal Global Outlook for Ice and Snow (2007). Original map design and cartography by Hugo Ahlenius; http://maps.grida.no/go/graphic/the-cryosphere-world-map.)

The Cryosphere

- Snow
- Sea ice
- Ice shelves
- Ice sheets
- Glaciers and ice caps
- Permafrost (continuous)
- Permafrost (discontinuous)
- Permafrost (isolated)

CRYSTAL STRUCTURE

Water—familiar, household H_2O—has a simple molecular arrangement, but this simplicity and familiarity disguise the fact that water is a rather peculiar substance. Hydrogen atoms within a water molecule are held to the oxygen atom by strong covalent bonds. The hydrogen atoms are grouped together on one side of the oxygen atom with a bond angle of 104.5°. Two electron pairs sit on the other side of the oxygen. This structure gives water molecules a strong polarity, with a positive charge on the side with the hydrogen atoms and a negative charge opposite to this, associated with the electron pairs. This dipolar nature creates strong intermolecular bonds between water molecules, as hydrogen atoms are attracted to the electron pairs of adjacent molecules. The resulting intermolecular *hydrogen bonds* are even stronger as a result of water's small molecular size, which allows close packing.

Water molecules group in a tetrahedral form, which should produce bond angles of 109.5°. The strong repulsion between the electron pairs distorts this, producing a lower bond angle of 104.5° in the liquid or vapor phase and giving water molecules a "bent" shape. In the solid phase, ice crystals form from individual water molecules bonded in symmetric, hexagonal plate structures (figure 2.1).

A number of different ice-crystal structures have been experimentally identified, but hexagonal ice (I_h) is the only structure that forms at the range of temperatures

a.

b.

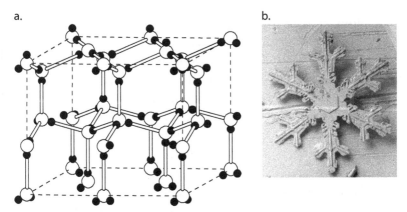

Figure 2.1. (a) Hexagonal symmetry and tetrahedral structures of the ice crystal lattice. Each oxygen atom (white sphere) is joined with four other oxygen atoms through covalent bonds with hydrogen atoms (black spheres). (b) Stellar dendrite snowflake, 0.25 mm, as imaged by a low-temperature scanning electron microscope. (Microphotograph courtesy of William Wergin and Eric Erbe, Beltsville Agricultural Research Center, Beltsville, Maryland.)

and pressures relevant to Earth's climate. Cubic ice (I_c) is found in ice crystals in the extreme cold temperatures of the upper atmosphere, where it can form at temperatures below 150 K. This structure is also expected at the low temperatures felt elsewhere in the solar system. More exotic, high-density ice structures have been experimentally produced at high pressures, but these are not found naturally on Earth.

The hexagonal symmetry of ice crystals results from the tetrahedral bonds of H_2O as water molecules freeze into a crystal lattice (figure 2.1a). In the solid phase, each hydrogen atom is still shared with an adjacent oxygen atom via hydrogen bonds, but the crystal lattice structure opens up to the tetrahedral bond angle of 109.5°. This results in an open, hexagonal planar structure.

Snowflakes in the atmosphere are not usually initiated through spontaneous (homogeneous) nucleation. Supercooled water droplets exist to temperatures as low as −40°C, as the low vapor density and continual movement of the air makes it difficult for ice crystals to nucleate (compared with, e.g., lake ice that forms in water). At subfreezing temperatures, water in clouds consists of a mixture of vapor, supercooled droplets, and ice crystals, something known as mixed clouds. A surface for nucleation, such as pollen, dust, or another ice crystal, helps to seed ice-crystal growth. Where present, such cloud condensation nuclei provide a surface for condensation as well as deposition and greatly facilitate cloud development.

Once nucleated, ice crystals grow in clouds through vapor deposition, as well as through collision and coagulation with other ice crystals. Because of curvature effects, the saturation vapor pressure over a plate-like ice surface is lower than that at the surface of a spherical liquid water droplet. This creates a favorable vapor pressure gradient that allows nascent ice crystals to outcompete water droplets with respect to vapor diffusion, and snowflakes grow at the expense of supercooled water droplets in a cloud. Once a snowflake become massive enough, it drifts to the ground, commonly experiencing modification en route through collisions or partial melting. These processes often cause snowflakes that reach the ground to be rounded or fragmented, rather than the textbook dendritic, stellar crystals of figure 2.1b.

DENSITY OF SNOW AND ICE

The geometric arrangement of this lattice structure is spacious, giving water the most unusual property of having a solid phase that is less dense than its liquid phase. At 0°C, water has a density of 1000 kg m^{-3}, whereas pure ice (I_h) has a density of 917 kg m^{-3}. Ice floats in its own melt, one of few substances to do so. Diamonds, germanium, gallium, and bismuth, all structurally similar to ice, also float in their own liquid. Imagine the sight of sunlight sparkling off of a diamond-berg in a sea of liquid diamond! But this is not to be found at Earth's surface temperatures and pressures.

Figure 2.2 plots the density of pure ice and water as a function of temperature. This plot also illustrates the unusual density inversion for freshwater. Pure water has its maximum density at 4°C, and it becomes less dense as it cools below this. The reason for this is not fully understood, but it is related to the angle of the hydrogen bonds in the rigid, low-density crystal lattice that characterizes water in its solid phase. To quote James Trefil, "water never quite forgets that it was once ice."

Water density continues to decrease below 0°C in supercooled water droplets (figure 2.2a). This density inversion is specific to freshwater. Salt content in seawater makes it more dense: 1028 kg m^{-3} for surface water with a temperature of 0°C and a salinity of 35 ppt. Dissolved ions in salt water also interfere with the molecular packing of

15

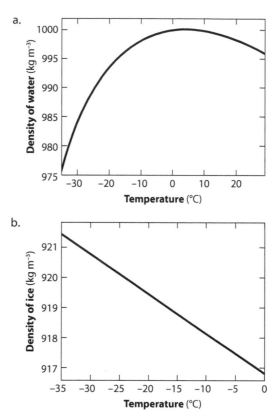

Figure 2.2. Density of (a) pure water and (b) ice as a function of temperature at mean sea level pressure. Maximum freshwater density is at 4°C.

water molecules, causing it to behave more like a regular liquid. Where salinity exceeds 24.7 ppt, density increases continuously as temperature drops to the freezing point. This is the case in most of the world's oceans. Seawater with a salinity of 35 ppt freezes at –1.9°C.

Once frozen, ice behaves like most solids, with increasing density as temperature drops; $\rho_i = 920$ kg m^{-3} at $-23°$C (figure 2.2b) and values reach 922 kg m^{-3} for the coldest ice to be found, in the Antarctic ice sheet. Ice density also increases slightly with pressure. The volume compressibility is about 1.2×10^{-10} Pa^{-1}, giving a density of 921 kg m^{-3} under the load of 4 km of ice, which is typical of the East Antarctic plateau.

The density of snow is less than that of crystalline ice. Snowflakes are composed of an arrangement of ice crystals, but snow that accumulates on the ground is a porous medium dominated by air pockets. Fresh, dry snow has an average density of about 100 kg m^{-3}, but this ranges from 20 to 200 kg m^{-3} or more, depending on the temperature, wind, and liquid water content during deposition. Densification occurs as snow settles and compacts, with dry snow densities increasing to as much as 400 kg m^{-3} in seasonal snowpacks that are subject to strong wind packing. Once the melt season begins, or in snow that becomes saturated from a rain event, further packing and sintering of grains increases the density to values of ca. 500 kg m^{-3}. Spherical packing models (which assume identical, individual spherical grains and air-filled pore space) predict a maximum density of 550 kg m^{-3}. Liquid water or refrozen ice in the pore spaces can further increase the bulk density.

Figure 2.3 illustrates this for a database of average snowpack densities, plotted as a function of day of year from a multiyear record of snow pits at Haig Glacier in the Canadian Rockies. Data are from three sites on the glacier and one site in the glacier forefield. Snow depths

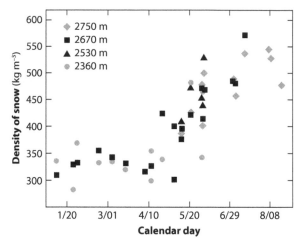

Figure 2.3. Average density of the snowpack as a function of day of year from four sites at Haig Glacier in the Canadian Rockies (50.7° N, 115.3° W). The upper three sites are on the glacier, and the lower site is from the glacier forefield.

in this compilation vary from 0.55 to 4.32 m (150 to 2160 mm water equivalent). The evolution of snowpack density is similar at all sites; fresh snow settles to densities of 300–350 kg m^{-3}, where it remains through the winter months, and a sharp increase in density accompanies warmer conditions and the onset of melting in May.

In the accumulation area of a glacier, snow that survives the summer melt season undergoes a gradual transition to *firn* and then glacial ice. This transition is accompanied by ongoing densification. The process is continual, and there is no clear distinction between snow and firn. In mountain glaciers, firn is often defined as snow that has lasted at least one melt season. This description does not apply to the polar ice sheets, however, where firn forms in the absence of melting. Firn can be loosely defined as dense, multiyear snow, with typical densities of 550 to 830 kg m^{-3}. The upper limit is well defined. This demarcates the transition from firn to ice, associated with closure of the pore space. In coastal mountain environments, where temperatures and snow accumulation rates are high, the transition from firn to ice takes place in a matter of years. This requires about 2500 years on the cold, dry plateau of East Antarctica.

The density of river, lake, and sea ice is similar to that of glacial ice during formation, with little or no air-filled void space. There can, however, be liquid water pockets, and sea ice contains brine pockets that give it a bulk salinity that is intermediate between the salinities of seawater and freshwater ice; 5–10 ppt is typical for first-year sea ice. There are also solid salt precipitates in sea ice. These inclusions can elevate the density, with measured values of up to 940 kg m^{-3}. As freshwater and sea-ice melt, vertical drainage channels form, absorbed shortwave radiation can melt subsurface ice, and the ice can deteriorate or become "rotten," with macroporous air pockets and channels. Ice densities during decay can drop to 700 kg m^{-3} or less in surface ice.

THERMODYNAMIC PROPERTIES OF SNOW AND ICE

The thermal characteristics of snow and ice play an important role in determining exchanges of energy

Table 2.1

Physical Properties of Snow and Ice at 0°C

Property	Freshwater	Ice	Sea Ice[a]	Snow[b] Fresh	Settled/Wet	Firn
Density (kg m^{-3})	1000	917	720–940	20–150	250–550	550–830
Albedo	0.1	0.1–0.6	0.1–0.7	0.8–0.9	0.4–0.6	0.3–0.4
Thermal conductivity (W m^{-1} °C^{-1})	0.56	2.11	1.91	0.03–0.06	0.1–0.7	0.7–1.5
Heat capacity (J g^{-1} °C^{-1})	4.218	2.12	2.12	2.09	2.09	2.09
Thermal diffusivity (m^2 s^{-1})	0.13	1.09	1.06	0.22	0.30	0.69
Latent heat of fusion (J g^{-1})	334					
Latent heat of sublimation (J g^{-1})	2834					

[a]Thermal properties for a temperature of –2°C, a salinity of 5 ppt, and a density of 850 kg m^{-3}.

[b]Thermal diffusivities calculated for densities of 100, 400, and 700 kg m^{-3}.

between the cryosphere, ocean, and atmosphere. They also govern the sensitivity of the cryosphere to climate change and many of the cryospheric feedbacks to this change. Select thermodynamic properties of snow and ice are compiled in table 2.1.

Specific and Latent Energy

One of the special properties of water is its resistance to phase changes. It requires 334 J to melt 1 g of ice at 0°C, an enormous amount of energy. Of common substances, only ammonia has a higher specific enthalpy (latent heat) of fusion. Direct conversion of ice from the solid to vapor phase is an even greater energy sink; sublimation of ice at 0°C requires 2834 J g^{-1}. Energy that is consumed for evaporation or sublimation is unavailable for melting snow and ice, a process that limits ablation in arid environments (e.g., high elevations in the tropical mountains and on the East Antarctic plateau).

The high specific enthalpies of phase change in water are a result of its small molecular size and strong hydrogen bonds, which reduce the intermolecular distance. A water molecule also forges hydrogen bonds with four adjacent water molecules, which distinguishes it from other compounds that have hydrogen bonds (e.g., methanol). The strong intermolecular bonds and stable tetrahedral structure make it difficult to separate water molecules. This also gives water unusually high freezing and boiling point temperatures.

Warming or cooling of liquid water also involves a great deal of energy. The specific heat capacity of water is 4.218 J g^{-1} °C^{-1} at 0°C and standard atmospheric pressure: once again, second only to ammonia among commonly found liquids. The heat capacity of ice is about half of this; snow and ice have less thermal inertia than that of liquid water, but they still require a lot of energy to warm or cool. Literature values of 2.05–2.13 J g^{-1} °C^{-1} are recommended for ice at 0°C. The heat capacity of pure ice increases slightly with temperature, an effect that can be approximated through a linear parameterization,

$$c_i = 2.115 + 0.008T \text{ J g}^{-1} °C^{-1}, \tag{2.1}$$

....

21

where temperature is in degrees Celsius. Cuffey and Paterson (2010) recommend an exponential relation,

$$k_i(T) = 2.072 \exp(-0.0057T) \text{ W m}^{-1}\,{}^{\circ}\text{C}^{-1}, \tag{2.5}$$

with temperature once again in degrees Celsius.

Within snow there are also temperature effects on thermal conductivity (as snow is composed of ice crystals), as well as complex influences from snow texture and microstructure. There is no single value for the thermal conductivity of snow, but the dominant influence on this parameter is the bulk porosity or density. Sturm et al. (1997) review several different parameterizations in the literature and recommend the following empirical relation:

$$k_s = \begin{cases} 0.023+0.234\rho_s, & \rho_s<0.156 \text{ g cm}^{-3}, \\ 0.138-1.01\rho_s+3.233\rho_s^2, & 0.156\leq\rho_s<0.6 \text{ g cm}^{-3} \end{cases}, \tag{2.6}$$

where snow density has units of grams per cubic centimeter. This gives a range from 0.03 to 0.70 W m$^{-1}\,{}^{\circ}$C^{-1} for densities typical of seasonal snow. Fresh, dry snow has much lower values (table 2.1), whereas rounded grains, dense wind slab, and liquid water content promote dense snow and thermal conductivities that are similar to that of water. As snow transforms from firn to ice, thermal conductivity continues to increase to a value that resembles that of other solids, such as mineral soils and rocks.

In sea ice, brine content introduces additional complications to heat transfer. Brine pockets decrease the thermal conductivity. There are once again numerous parameterizations in the literature, commonly of the form

$$k_{si} = \rho_{si}/\rho_i(k_i + \beta S/T),$$ (2.7)

where S and T are the salinity and temperature of the ice, in ppt (‰) and degrees Celsius, ρ_i is the density of pure ice, and ρ_{si} is the density of the sea ice. The density ratio is introduced to capture the influence of air-filled pore space, which reduces the thermal conductivity relative to pure ice. Literature values of β are in the range 0.09–0.12 W m^{-1} ppt^{-1}. For a case with $T = -2°C$ and $S = 10$ ppt, $\beta = 0.09$ W m^{-1}ppt^{-1} gives $k_{si} = 1.68\rho_{si}/\rho_i$ W m^{-1}°C^{-1}. At $-10°C$ and $S = 10$ ppt, this increases to $k_{si} = 2.13\rho_{si}/\rho_i$ W m^{-1}°C^{-1}.

Thermal conductivity is a primary parameter for heat transfer, but thermal diffusion also depends on the density and specific heat capacity of the material. The thermal diffusivity, $\kappa = k/(\rho c_p)$, determines the depth and rate of penetration of atmospheric temperature signals into snow and ice. Typical values are given in table 2.1. Daily temperature signals (diurnal warming and cooling) penetrate to tens of centimeters in seasonal snowpacks. Diffusion of the annual temperature cycle in glaciers, ice sheets, and permafrost reaches a depth of approximately 10 m.

Heat transfer into snow and ice is not a purely diffusive phenomenon. There can also be heat advection from wind-pumping (ventilation), from rainfall and meltwater that percolate into the snowpack, and from vapor diffusion in the snowpack, driven by vertical gradients in vapor pressure. In floating ice and soils, capillary pressures can also cause water to be wicked upward. Liquid water that percolates into a cold snowpack will refreeze, and vapor moving through the snowpack can be deposited. Both of these mass-transfer processes release latent heat, warming the snowpack. This is an important part of the ripening process in glacial environments: It can give mean annual surface temperatures several degrees Celsius warmer than the mean annual air temperature.

SNOW ALBEDO

The influence of snow and ice in the climate system stems largely from their high albedo. Shortwave radiation is backscattered from snow and ice, with the reflectivity strongly dependent on wavelength. Neglecting this complexity for now, the broadband surface albedo can be defined from the ratio of reflected to incoming shortwave radiation, $\alpha = Q_S/Q_s$.

A range of albedo values for snow and ice is given in table 2.1. Albedo is highly variable, spatially and temporally, so it is not recommended to adopt a single, constant value for modeling of snow and ice in the climate system. As an illustration, figure 2.4 plots the evolution of surface albedo through the melt season as measured on Haig Glacier. Fresh snow has an albedo of 0.8–0.9. This is typical of seasonal snowpacks during winter months and year-round values in the accumulation area of the polar ice sheets. Snow albedo typically decreases in old snowpacks, particularly when temperatures are above 0°C and meltwater is introduced into the snow. Albedo values of a mature, wet snowpack are closer to 0.6 and can fall well below this (figure 2.4).

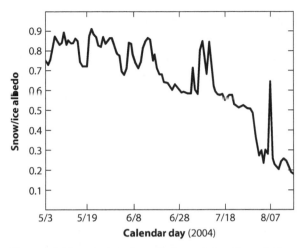

Figure 2.4. Measured surface albedo evolution through the melt season at Haig Glacier in the Canadian Rockies. The transition from seasonal snow cover to bare glacial ice is evident from the albedo drop in early August. High-albedo spikes in the record are associated with snowfall events.

Several processes are responsible. Snow-grain metamorphism generally causes an increase in crystal size in the days and weeks after a fresh snowfall. The effective optical radius of spherical snow particles is ca. 50 μm for fresh snow, increasing to 100 μm within days and 1 mm or more in mature, melting snowpacks. This increase in grain size increases the path length for solar radiation transmittance in the near-surface snow layer, effectively reducing the incidence of scattering reactions at intergranular snow–air interfaces and reducing the snow albedo.

There are compounding effects beyond just grain metamorphism in melting snowpacks. Chemical impurities and meltwater content also reduce snow albedo. It requires only a few parts per million of impurities to cause an albedo reduction of several percent. The effect depends on the type of impurity, with black carbon (soot) inducing a discernible impact for concentrations of order 0.1 ppm, comparable with mineral dust levels of 10 ppm. Natural snowpacks in tropical and midlatitude areas have impurity concentrations of this order of magnitude or greater, but dust concentrations in polar snowpacks are commonly less than 0.1 ppm.

The effects of grain-scale recrystallization may saturate after several days or weeks, but meltwater effects and impurity concentration increase throughout the melt season. The result is a reduction in albedo to late-summer values as low as 0.3 in midlatitude snowpacks. Values vary between sites and from place to place in a given snowpack, as a function of the concentration and type of impurity. The role of liquid water is unclear in scattering reactions; because the refractive index of snow crystals and liquid water is similar, there should be little effect. However, water content may increase the effective radius of grains. On a macroscale, ponded surface water causes significant albedo reductions on snow and ice surfaces, accelerating spring and summer melt on lake ice, sea ice, and glaciers.

Ice typically has a lower albedo than snow, but estimates in the literature again vary substantially, from 0.1 to 0.6. A value $\alpha_i = 0.2$ is typical of midlatitude glaciers, whereas $\alpha_i = 0.5$ is more representative of sea ice and the ablation zones of polar glaciers and ice sheets. Crystal size, impurity concentration on the glacier or sea-ice surface, liquid water, and superimposed ice content again play large roles, which gives rise to a large amount of spatial variability; micro-topography causes some areas to pond water and debris, whereas other areas are flushed. On glacier surfaces, albedo generally decreases at lower elevations in the ablation zone due to higher melt rates and longer exposure times for the surface. These influences contribute to higher debris concentrations within a given melt season and cumulatively, over many years. The albedo of debris-rich ice is 0.1–0.2. Clean ice has values closer to 0.4. Superimposed (refrozen) ice and blue ice are much brighter, with values closer to 0.6. Blue ice is found in the ablation area of polar ice caps, where wind scour can create bare-ice zones.

Albedo values for lake and sea ice depend on the age, type, and thickness of the ice. Young ice that is only a few centimeters thick is highly transparent to solar radiation, so absorption in the underlying water reduces the effective spectral albedo to values of 0.1–0.2. Young "gray ice" that is up to 30 cm thick is still relatively dark, with reported albedo values of 0.3–0.4. Ice becomes opaque as it thickens beyond this, and there is also more complete ice cover in an area (versus a large fraction of open water for young, growing ice), so albedo values of developed first-year or multiyear ice are in the range 0.6–0.7, dropping during the melt season in the presence of melt ponds. When sea ice has a snow cover exceeding a few centimeters, snow spectral properties dominate the albedo, so values of 0.8–0.9 are typical.

3 SNOW AND ICE THERMODYNAMICS

The world has signed a pact . . .
It is a covenant to which every thing,
even every hydrogen atom, is bound.
The terms are clear: if you want
to live, you have to die . . .
The world came into being
with the signing of a contract.
A scientist calls it the Second Law
of Thermodynamics.
—Annie Dillard, *Pilgrim at Tinker Creek*

THE MATERIAL PROPERTIES OF ICE DISCUSSED IN CHAPTER 2 govern its macroscale behavior in the global cryosphere. Much of the cryosphere's influence in the climate system stems from its role in local, regional, and global energy balance. Snow and ice reflect solar radiation back to the atmosphere, reducing heat absorbed at Earth's surface. The latent heat associated with phase changes provides a thermal inertia that is similar to the moderation of maritime climates from the heat capacity of the ocean. In the autumn and winter, freezing of water in rivers, lakes, and the oceans releases energy and provides a source of sensible heat. Melting of snow and ice in the spring and summer has the opposite influence, providing an effective "heat sink" for the radiative and sensible heat that would otherwise warm a region. These processes all involve exchanges of energy between the atmosphere and the snow or ice at Earth's surface, as well as the internal thermal evolution of a snow or ice mass. The underlying physics are relevant to all of the elements of the cryosphere, so we consider the main aspects of snow and ice thermodynamics in this chapter. Features of snow and ice thermodynamics that are specific to a particular element of the cryosphere are deferred to subsequent chapters.

ENERGY BALANCE AT THE SNOW/ICE SURFACE

The surface energy balance of melting snow and ice has been studied extensively, although there are still some unresolved questions. The first challenge is to decide whether a snow or ice surface is two- or three-dimensional; that is, is it a surface or a volume? It is both in some ways. There is a two-way exchange of energy between the atmosphere and surface, through several different processes of energy transfer, and this determines the net energy that is supplied to or lost from snow or ice during a given time interval. At the same time, some of the energy fluxes at the surface, such as transmission and reflection of shortwave (solar) radiation, involve the upper several centimeters or decimeters of the snowpack or the ice. Turbulent eddies that are responsible for sensible heat flux (defined later) also transport energy

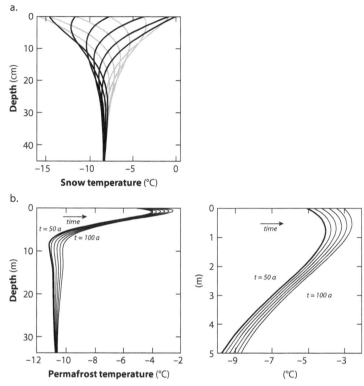

Figure 3.2. (a) Modeled penetration of an idealized diurnal temperature cycle in a seasonal snowpack. Temperature is shown for the upper 45 cm of a 1-m snowpack, plotted at 2-hour intervals for a 24-hour period. Black lines indicate a warming from 0200 until 1200 (left to right), and gray lines correspond with a cooling phase from 1400 until 0000 (right to left). (b) Temperature response to a surface warming in the upper 35 m of a 100-m-deep section of permafrost. This shows the diffusion of a surface temperature warming to depth for a 2°C temperature increase across 50 years. The temperature profile is plotted every 10 years (50, 60 . . . 100 years, from left to right) for the numerical experiment described in the text. The inset provides a closer view of the upper 5 m.

The same physics applies to the permafrost in figure 3.2b, but the temporal and spatial scales of this example are much greater. Figure 3.2b plots the solution of Eq. (3.5) in the upper part of a 100-m-deep section of permafrost. This example includes both "equilibrium" and transient surface temperature boundary conditions to illustrate the diffusion of a surface warming trend in permafrost. The simulation is based on an idealized sinusoidal annual temperature cycle that repeats itself across 1000 years. The deep permafrost equilibrates by the end of the 1000-year "spin-up" simulation. The mean annual surface temperature for this initialization is –11°C, with a geothermal heat flux of 0.05 W m^{-2} prescribed at the base. At depth, permafrost temperatures reach a steady state after several centuries. The upper 10–15 m of the ground feels the annual temperature cycle, as seen in the initial temperature profile (heavy line labeled $t = 50a$ in figure 3.2b).

After this spin-up, the reference annual temperature cycle is held fixed for the first 50 years of a model experiment, followed by the imposition of a linear warming trend of 2°C during the next 50 years. Temperature profiles in figure 3.2b are plotted every 10 years for this 100-year simulation. These profiles are snapshots from October 1 of each new decade. The cold wave from the prior winter is seen at about 8 m depth, and the warm wave from the recent summer is visible at 1 m depth. Temperatures throughout the upper 40 m of the ground increase as surface temperatures rise in the second half of the simulation, with attenuation of the warming signal with depth. Below 35- to 40-m depth, the surface temperature change is not yet detectable.

This indicates the timescale of temperature response to climate change in frozen ground, although the detailed response depends on the thermal conductivity, heat capacity, density, and porosity of the soil or rock matrix. Temperature diffusion in pure ice has a similar timescale to that of frozen ground.

Radiative fluxes at the atmospheric interface are of order hundreds of watts per square meter. The geothermal heat flux from the ground into the base of a snowpack or a terrestrial ice body (glaciers, ice sheets, and permafrost) is negligible compared with this, with typical values of 0.04–0.06 W m^{-2}. Although this is a minor component of the overall energy budget, ground heat flux still helps to warm the base of a seasonal snowpack over the course of the winter, particularly in deep snowpacks where basal snow is well insulated from the atmosphere. Similarly, geothermal heat flux provides a steady trickle of heat to the base of terrestrial ice masses that are tens to thousands of meters thick. This ultimately limits the depth of permafrost, and for many of the world's glaciers and ice sheets, this supply of heat drives basal melting at rates of several millimeters per year.

The basal boundary is more complex and important for sea ice and where glaciers and ice sheets are in contact with the ocean, as oceanic heat fluxes are commonly several watts per square meter, greatly exceeding geothermal heat. Even in sea ice, however, the primary concern in modeling snow/ice thermal evolution and melt is quantification of energy fluxes at the surface–atmosphere interface. The next section describes these fluxes in greater detail.

SURFACE ENERGY BALANCE PROCESSES

Here we take a brief look at the different terms in the surface energy budget. These processes are the subject of many excellent textbooks and research papers; this section is limited to a brief overview of some important considerations for the cryosphere.

Shortwave Radiation

Shortwave radiation is the main driver of snow and ice melt in most environments. Incoming solar radiation is absorbed and scattered as it traverses the gauntlet of atmospheric gases and aerosols (suspended particles such as water droplets, ice crystals, and dust). The processes of absorption and scattering depend on the wavelength of the electromagnetic radiation and the size of the obstacle (gas or aerosol). Similarly, backscatter (reflection) from ice crystals is also wavelength dependent, as discussed in chapter 2. This complexity is commonly neglected in cryosphere studies, however, and the shortwave radiation and albedo in (3.1) are defined based on an integrated broadband spectrum, from about 0.2 to 2.5 μm. Alternatively, shortwave radiation can be modeled in two or three wavelength bands, for instance ultraviolet, visible, and near-infrared frequencies. This is becoming more common in energy balance studies as it permits a separate treatment of visible and near-infrared albedos, which differ markedly for snow (figure 2.5).

The average top-of-atmosphere solar irradiance is the solar constant, I_0, which is approximately equal to 1366 W m^{-2}. The solar constant is not really constant; daily and decadal variations of up to a few watts per square meter are common, as a result of variable solar convective activity. The instantaneous top-of-atmosphere solar irradiance also differs from the solar constant as a function of the Earth–Sun distance. This instantaneous distance is denoted R. Top-of-atmosphere irradiance, Q_{S0}, is calculated from

$$Q_{S0} = I_0 \left(\frac{R_0}{R} \right)^2, \tag{3.8}$$

where $R_0 = 1.5 \times 10^8$ km is the mean Earth–Sun distance. Peak top-of-atmosphere solar irradiance occurs when the Earth is at its closest approach to the Sun, a time known as the perihelion. Perihelion currently occurs on January 3, although its timing varies on timescales of 10^4 years as part of the Milankovitch cycles (see chapter 9).

Incoming shortwave radiation at the surface, also known as insolation, is made up of two main components: direct and diffuse solar radiation. A third contribution, direct light that is reflected from the surrounding terrain, can also add to the surface insolation. Direct solar radiation is the radiative flux from the solar beam, which comes in at a zenith angle Z, measured from the normal to the geoid surface. The zenith angle is a function of latitude, time of year, and time of day. Potential direct (clear-sky) solar radiation on a horizontal surface can be estimated from

$$Q_{S\phi} = Q_{S0} \cos(Z) \psi^{P/[P_0 \cos(Z)]}, \tag{3.9}$$

where ψ is the atmospheric transmissivity at sea level, P is the air pressure at the site, and $P_0 = 101.325$ kPa is the mean air pressure at sea level. $P/P_0 \cos(Z)$ in (3.9) account for the effects of atmospheric attenuation due to the amount of atmosphere that the direct beam must traverse, a function of both elevation (atmospheric pressure) and slant path.

It is common to use the clear-sky atmospheric transmissivity in (3.9), $\psi_0 \approx 0.84$. The effects of cloud cover or varying atmospheric absorption (e.g., associated with dust or aerosols) can also be incorporated in ψ. In thick haze or smog, $\psi \approx 0.6$, and under heavy cloud cover, $\psi \to 0$. The actual direct solar radiation at a site only equals the potential direct solar radiation under clear-sky conditions.

In mountain topography, the effects of surface slope, aspect, and shading also need to be incorporated in estimates of potential direct solar radiation. This can be approximated from digital elevation models, and the solar geometry can be modeled for a particular location, time of year, and time of day, allowing detailed calculations of potential direct solar radiation at any point in space.

In addition to the direct solar beam, diffuse radiation reaches a site from all directions in the sky hemisphere. Diffuse atmospheric radiation arises due to Rayleigh scattering off of atmospheric gases and Mie scattering off of aerosols, water droplets, and ice crystals. Illumination from diffuse light is the reason that there is not complete darkness when the Sun is obscured. Diffuse atmospheric radiation is close to isotropic (derived equally from all points in the sky vault) when it is overcast, but it is generally anisotropic, with more radiation in proximity to the direct beam. Total incoming solar radiation at the surface, $Q_S^{\downarrow}$, is equal to the sum of the direct, diffuse, and terrain contributions. This is often called the global radiation, although this is unfortunate terminology as it is not global in the true sense of the word.

Longwave Radiation

Longwave radiation is also known as infrared, thermal, or terrestrial radiation. It is electromagnetic energy in the spectral band from roughly 3 to 100 μm. Earth surface temperatures produce emissions in this range, with peak terrestrial radiation occurring at a wavelength of about 10 μm. Spectrally integrated longwave radiation can be estimated from the Stefan–Boltzmann equation, $Q_L = \varepsilon \sigma T^4$, where ε is the thermal emissivity, σ is the Stefan–Boltzmann constant, 5.67×10^{-8} W m^{-2} K^{-4}, and T is the absolute temperature of the emitting surface. By definition, ε is the ratio of emitted longwave radiation to that which would be emitted by a perfect blackbody (a perfect emitter or absorber); $\varepsilon = 1$ for a perfect blackbody.

Snow emits as a near-perfect blackbody at infrared wavelengths, with an emissivity ε_s of 0.98–0.99. To good approximation, then,

$$Q_L^{\uparrow} = \varepsilon_s \sigma T_s^4, \tag{3.10}$$

for surface temperature T_s. This is a loss of energy from the snow or ice surface. For saturated snow and ice, $\varepsilon_s \to 1$ (a perfect blackbody), so for a melting snow or ice surface, $T_s = 273.15$ K and $Q_L^{\uparrow} \approx 315$ W m^{-2}.

Incoming longwave radiation is more variable and is difficult to predict without knowledge of lower-troposphere water vapor, cloud, and temperature profiles. We are all empirically familiar with the experience of warm, cloudy nights versus cold, clear nights; we are feeling the impact of variations in $Q_L^{\downarrow}$. The Stefan–Boltzmann equation still holds in the atmosphere, but the longwave flux to the surface comes from different heights (temperatures) in the atmosphere, and the air is made up of an ensemble of gases with different infrared emissivities. Water vapor and CO_2 are strong absorbers/emitters and are the dominant gases that influence $Q_L^{\downarrow}$, although other greenhouse gases and aerosols contribute. A spectrally and vertically integrated radiative transfer calculation is needed to rigorously predict the longwave radiation incident on the surface. In snow studies, $Q_L^{\downarrow}$ is usually parameterized at a site, assuming that the atmosphere emits longwave radiation with an effective emissivity, ε_a, such that

$$Q_L^{\downarrow} = \varepsilon_a \sigma T_a^4, \tag{3.11}$$

where T_a is the near-surface air temperature. Various parameterizations of ε_a have been proposed, typically as a function of atmospheric humidity and cloud cover.

In many terrestrial environments, vegetation creates another potential longwave heat source. Trees in particular can provide local sources of both thermal radiation and sensible heat; the tree wells that surround tree trunks attest to this. In coastal mountain settings such as western North America and Norway, such tree wells can be several meters deep. For valley glaciers, side walls that heat up in the summer sun can also provide a significant source of longwave radiation.

Subsurface Energy Fluxes

Subsurface energy fluxes, denoted Q_G, are flows of energy between the surface and the underlying snow or ice. These arise due to thermal diffusion and penetration of shortwave radiation. These processes are discussed above with respect to the internal temperature distribution in the snowpack. For the surface energy balance of concern here, these exchanges represent a flow of energy either toward or away from the snow/ice surface.

Diffusive energy exchange between the surface and subsurface occurs in the presence of a temperature gradient as per (3.3). Conductive heat flux to the surface, Q_{Gc}, is positive when $\partial T/\partial z < 0$. The sign of Q_{Gc} is variable and the magnitude is usually small: $|Q_{Gc}| <$ 10 W m^{-2}. There is commonly a radiatively driven diurnal cycle in Q_{Gc} associated with diffusion of atmospheric temperature variations to depths of a few decimeters. In isothermal snow or ice, $\partial T/\partial z = 0$, giving $Q_{Gc} = 0$. However, there can still be diurnal cycles in the near-surface, driven by overnight longwave cooling. This does not occur if free water is present at the surface, as energy losses drive refreezing. The associated latent heat release provides a thermal buffer that helps to maintains surface temperatures near 0°C.

Transmitted shortwave radiation, Q_{GS}, warms or melts the underlying snow or ice. Because longwave radiation losses induce surface cooling, it is not unusual to find a frozen surface layer overlying subsurface meltwater pockets in ice. Q_{GS} represents a reduction in the absorbed shortwave radiation that is available for surface melt, $Q_S^{\downarrow}$ $(1 - \alpha_s)$, but it is often convenient to count this within Q_G: $Q_G = Q_{Gc} - Q_{GS}$. There can also be sensible heat transport by meltwater percolation into a snowpack, but this is rarely measured or modeled.

Turbulent Heat Fluxes

Turbulent heat fluxes describe the exchange of energy between the snow surface and the near-surface atmospheric boundary layer. Friction at the surface–atmosphere interface creates dissipation of momentum through both direct (molecular) viscous drag and turbulent eddy motions. Direct viscous effects are confined to within a few centimeters of the surface and are generally minor; the energy fluxes in (3.1) are primarily associated with turbulent eddies. If the surface is rough and the wind is strong enough to create turbulence, sensible and latent energy at the surface, Q_H and Q_E, are created as a by-product of the dissipation of momentum and kinetic energy in the atmospheric flow over the snow or ice.

The equations used to estimate turbulent energy fluxes are a complex mixture of theory and empiricism. The simplest approach is to assume a well-mixed atmospheric boundary layer, with vertical fluxes of sensible and latent energy proportional to the wind speed. This can be parameterized through

$$Q_H = \rho_a c_{pa} C_H u_a (\theta_a - \theta_s),$$
$$Q_E = \rho_a L_{s/v} C_E u_a (q_v - q_s),$$
$$(3.12)$$

where ρ_a and c_{pa} are the density and specific heat capacity of the air, $L_{s/v}$ is the latent energy of sublimation or vaporization, u_a is the wind speed, and C_H and C_E are dimensionless bulk exchange parameters for heat and moisture. The atmospheric variables θ and q refer to the potential temperature and specific humidity of the air. Surface values of potential temperature and humidity in (3.12), sometimes referred to as "skin" values, are those within ca. 1 mm of the surface. These are taken to be the surface temperature of the snow or ice and the saturation specific humidity at this temperature.

Snow and ice surfaces are characterized by a stable boundary layer, with cold air near the surface and strong vertical gradients in temperature, humidity, and wind speed. Stability adjustments can be built into the aerodynamic coefficients of (3.12) or the bulk aerodynamic formulas can be modified to include more realistic assumptions about boundary-layer profiles. The latter approach introduces a different model for closure, variously known as Prandtl theory, flux-gradient theory, the profile method, scalar transfer theory, or the eddy-diffusivity model of turbulent fluxes. This treatment is conceptually different from the bulk aerodynamic approach of the slab mixed layer, although the two methods converge mathematically in some applications.

Based on assumptions about the vertical gradients of the thermodynamic variables in the near-surface boundary layer, Prandtl theory essentially parameterizes turbulent fluxes as a form of bulk diffusion, with eddy diffusivities K_H and K_E. There are direct parallels between these parameters and the idea of eddy viscosity that is used in oceanographic modeling. Eddy viscosity is a construct, unrelated to molecular viscosity. Similarly, the eddy diffusivities for heat and moisture resemble thermal and hydraulic diffusivities, but they are not material properties. To some degree it is reasonable to think of turbulent transfers as a form of diffusion: momentum, heat, and moisture are transferred from high to low concentrations, creating a more homogeneous lower boundary layer. The analogy with true diffusion has limitations. Turbulent exchange is not always effective at mixing the lower boundary layer, and the efficacy of mixing depends on surface roughness properties, wind strength, wind shear, and lower boundary layer stability. In principle, these influences can be incorporated in the eddy diffusivities, K_H and K_E.

According to classical Prandtl theory, wind speed increases with height above the surface following

$$\frac{\partial u_a}{\partial z} = \frac{u_*}{kz}. \tag{3.13}$$

Parameter k is von Kármán's constant, which has an empirically determined value of 0.4, and u_* is a characteristic velocity. This can be integrated to give the well-known log relationship for boundary layer winds,

$$u_a(z) = \frac{u_*}{k} \ln\left(\frac{z}{z_0}\right), \tag{3.14}$$

where z_0 is an integration constant and is known as the *surface roughness*. Mathematically, surface roughness length is defined as the height at which the wind speed goes to zero. Physically, the roughness length relates to the degree of mechanical coupling of the snow surface and the boundary-layer airflow. Rougher surfaces impose a greater viscous perturbation and are more likely to lead to turbulent eddies, given sufficient airflow. Aerodynamic roughness values are less than the geometric roughness elements, but z_0 is generally proportional to geometrical asperities and undulations in the surface. Published values of z_0 range from 0.1 mm to a few centimeters over snow and ice, based on wind profile measurements in neutral stability conditions. Sub-millimeter measurements refer primarily to grain-scale roughness elements. Values of 1–10 mm are typical of melting surfaces, which can develop sun cups and relatively large-scale surface undulations.

Working from the velocity profile in (3.14) and with similar assumptions for the profiles of potential temperature and specific humidity,

$$
\begin{aligned}
Q_H &= \rho_a c_{pa} K_H \frac{\partial \theta_a}{\partial z} = \rho_a c_{pa} k^2 u_a \left[\frac{\theta_a(z) - \theta_a(z_{0H})}{\ln(z/z_0)\ln(z/z_{0H})} \right], \\
Q_E &= \rho_a L_{s/v} K_E \frac{\partial q_v}{\partial z} = \rho_a L_{s/v} k^2 u_a \left[\frac{q_v(z) - q_v(z_{0E})}{\ln(z/z_0)\ln(z/z_{0E})} \right].
\end{aligned}
\tag{3.15}
$$

Here, z_{0H} and z_{0E} are the roughness length scales for sensible and latent heat fluxes. These are distinct from the roughness length in turbulent momentum exchange. By analogy, however, they can be taken as the heights at which θ_a and q_v are equal to the surface (skin) values. Implicit in (3.15) is the assumption that the eddy diffusivities for momentum, sensible heat, and latent heat transport are equal. This expression also assumes neutral stability in the lower boundary layer. Equation (3.15) can be adjusted to parameterize the effects of atmospheric stability, which will amplify or limit the extent of turbulent energy exchange.

FIELD EXAMPLE OF SURFACE ENERGY BALANCE

As an illustration of the different energy balance fluxes over snow and ice surface, figure 3.3 plots time series of temperature, humidity, radiation fluxes, and turbulent energy fluxes for the period July 15 to August 15, 2008, on Kwadacha Glacier in the northern Rocky Mountains (57°50′ N, 124°57′ W). Meteorological conditions and the four components of radiation are measured from an automatic weather station (AWS) at 2000 m altitude on the glacier. Turbulent fluxes are calculated from (3.15), with a surface roughness value of $z_0 = 1$ mm and $z_{0H} = z_{0E} = z_0/100$. The AWS was set up on a tripod on the winter snow surface in May, and by August 6 the seasonal snowpack had melted to expose bare glacier ice at the AWS site. These data therefore capture the transition

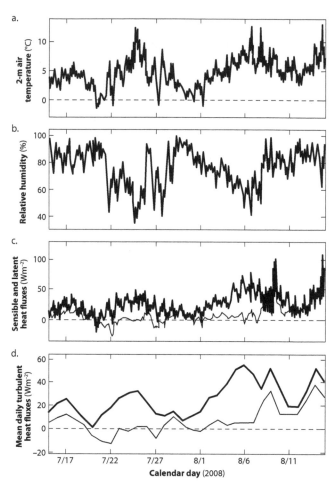

Figure 3.3. Two-meter meteorological fields and surface energy balance fluxes on Kwadacha Glacier, July 15 to August 15, 2008. All fields are based on 30-minute automatic weather station data. (a) Air temperature. (b) Relative humidity. (c, d) Sensible heat flux (heavy black lines) and latent heat flux (thin gray lines), with (d) giving the mean daily fluxes. (e) Net shortwave radiation.

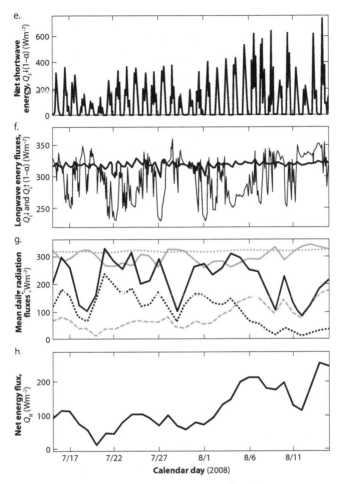

(f) Incoming (thin line) and outgoing (heavy line) longwave radiation. (g) Mean daily radiation fluxes. Incoming and outgoing longwave radiation (solid and dotted gray lines), incoming and reflected shortwave radiation (solid and dotted black lines), and net radiation flux (heavy gray dashed line). (h) Mean daily net energy flux.

from a melting snow surface to melting glacier ice, associated with an albedo reduction from 0.63 over the late-July snow surface to 0.22 over the ice (table 3.1). The observed melt at the site was 924 mm w.e. for the period July 15 to August 15, approximately half of the total summer melt.

Table 3.1 summarizes the mean energy fluxes during this period and for the full summer melt season. Net shortwave radiation drives the summer melt, accounting for an average daily energy supply of 105.4 W m^{-2} to the snow/ice surface from July 15 to August 15 (figure 3.3e, 3.3g). This increases dramatically when the ice is exposed and surface albedo drops. As a result, ice melt during the final 10 days is roughly equal to the snow melt from the preceding 21-day period, despite lower levels of incoming solar radiation. Longwave radiation acts as a net heat sink of −20 W m^{-2} during the 1-month period. Outgoing longwave emissions are relatively uniform from the melting surface (figure 3.3f, 3.3g), following Eq. (3.10). Incoming longwave flux falls to values below 250 W m^{-2} on clear nights, when humidity drops (figure 3.3b). The net radiation during the 1-month period is 85.3 W m^{-2}, 72% of the net energy, Q_N. For the full summer, net radiation makes up 78% of Q_N. Outside of the tropics, this is typical of the surface energy balance over snow and ice during the melt season, with 60% to 90% of available melt energy derived from shortwave radiation.

The turbulent heat fluxes make up the remaining fraction of the energy available for snow and ice melt. Sensible and latent fluxes both represent a net supply of

Table 3.1

Energy Balance Fluxes for Snow and Ice Surfaces on Kwadacha Glacier in the Canadian Rockies

Period (2008)	$Q_S^\downarrow$	$Q_S^\uparrow$	α_s	$Q_L^\downarrow$	$Q_L^\uparrow$	Q_H	Q_E	Q_N	T_a (°C)	q_{lv} (g/kg)	m_{mod} (mm w.e.)	m_{obs} (mm w.e.)
July 15–Aug. 15	214	108	0.51	299	319	26	7	119	4.0	5.24	982	924
July 15–Aug. 5	230	143	0.63	292	318	21	2	87	3.1	4.94	494	475
Aug. 6–Aug. 15	183	42	0.22	312	321	39	20	190	5.8	5.97	489	449
May 25–Aug. 20	245	154	0.61	290	317	19	−1	82	2.6	4.72	1861	1820

Note: All fluxes are averages for the period, measured in W m^{-2}, with symbols as defined in text. Modeled melt, m_{mod}, is based on Q_N and Eq. (3.2a), whereas observed melt, m_{obs}, is based on measured snow densities and an ultrasonic depth gauge.

energy to the glacier in this example, 26 and 7 W m^{-2}. These values are sensitive to the assumptions about surface roughness. For $z_0 = 0.1$ mm, the average fluxes decrease to 17 and 5 W m^{-2}, and $z_0 = 10$ mm gives 48 and 13 W m^{-2}. There is a persistently stable stratification in the near-surface boundary layer over snow and ice, so stability corrections typically reduce the magnitude of the turbulent fluxes. Such corrections may be warranted in this case in order to reduce the modeled melt in table 3.1 to match the observed melt. As this is just an example here, there has been no effort to tune the energy balance model.

Sensible heat flux is correlated with net radiation at Kwadacha Glacier, as air temperatures increase over the bare glacier ice in August. Shortwave and sensible heat fluxes act in concert to double the available melt energy in this period. Across the time period of figure 3.3, sensible heat flux was positive 98% of the time while latent heat was positive 76% of the time (figure 3.3e). This was a period of warm temperatures and moist air masses, with condensation prevailing over sublimation. For the full summer, sensible and latent heat fluxes were positive 90% and 52% of the time, with periods of sublimation giving a slightly negative mean latent heat flux (-1 W m^{-2}).

The energy fluxes in this example are from a specific example of a midlatitude mountain glacier, but they are representative of the energy fluxes that drive summer melting of seasonal snow, sea ice, river and lake ice, and the permafrost active layer. The terms in the energy balance look markedly different in the late autumn and winter, when seasonal snow and ice spread over the high-latitude oceans and continents. At this time, incoming shortwave radiation is low, with the sun winking out at polar latitudes ($Q_S^{\downarrow} = 0$). Longwave radiation continues to provide a net loss of energy, and the net energy, Q_N, is negative. The atmosphere, land, and surface waters cool, precipitation falls as snow, and on those parts of the planet that dip below 0°C, rivers, lakes, and ocean waters begin to freeze up. The next several chapters discuss the main features of these different elements of the cryosphere.

SUMMARY

This chapter has provided a brief, general overview of the essential thermodynamics that drive freezing, melting, and the internal temperature evolution of the elements of the cryosphere. Fluxes of energy between the atmosphere, cryosphere, and underlying substrate (ground or water) shape the extent and thermal character of snow and ice and also feed back on regional and global energy balances. Radiative, sensible, and latent heat exchanges between the atmosphere and cryosphere are all important, and chapter 8 describes the ways in which these exchanges impact weather and climate.

The internal heat transfer and surface energy balance processes and the examples described here apply to all aspects of the cryosphere. Subsequent chapters build on this through considerations that are relevant to specific aspects of freshwater and sea ice, glaciers, and permafrost. In particular, chapter 4 builds on the physics that define the internal temperature evolution, through coupling of Eq. (3.5) with the equations that describe lake ice growth. The additional physics of phase change and freezing front migration are also important for sea ice and permafrost growth, as all of these are driven by thermal diffusion of cold atmospheric temperatures to depth in water, snow, ice, or the ground. Similarly, changes in any aspect of the surface energy balance, through climate warming or through changes in humidity, winds, clouds, and so forth, will alter the energy that is available for snow and ice melt, via Eq. (3.1). This is central to cryosphere–climate interactions, and chapters 8 and 9 return to several of these processes.

4 SEASONAL SNOW AND FRESHWATER ICE

Come, see the north wind's masonry.
Out of an unseen quarry evermore
Furnished with tile, the fierce artificer
Curves his white bastions with projected roof
Round every windward stake, or tree, or door.
Speeding, the myriad-handed, his wild work
So fanciful, so savage, naught cares he
For number or proportion.
—Ralph Waldo Emerson, "The Snowstorm"

SNOW AND ICE TAKE ON A WONDROUS ARRAY OF MANI-festations, each with its own role to play and story to tell in terms of the global climate system. The next several chapters provide overviews of the geography and physics of the main elements of the cryosphere. We begin with the seasonal blanket of snow and ice that covers much of the land surface each winter.

SEASONAL SNOW

Outside of the tropics, almost all regions of the world experience seasonal snowfall. The first snow of the year is excitedly anticipated by many people each autumn. By midwinter, snow covers about half of the land mass in the Northern Hemisphere (see table 1.1), and by the time spring arrives, most midlatitude denizens are cursing the snow and anxiously awaiting its seasonal exodus from the landscape. This cycle is repeated year after year, with nostalgia for the beauty that snow imparts on the landscape and the recreational opportunities offered by snow and ice giving way to practical considerations such as the inconvenience that snowfall can bring to transportation, mobility, and comfort.

Snow in the Atmosphere

The physical properties of snow define and shape its influence on climate and society. Temperatures in the middle and upper troposphere are below 0°C, so ice crystals can be found everywhere in the global atmosphere. As discussed in chapter 2, mixed clouds below this temperature consist of a blend of water vapor, ice crystals, and supercooled water droplets.

Once nucleated, ice crystals grow through deposition of water vapor. Saturation vapor pressure on the surface of rounded water droplets is greater than that of ice crystals, due to the effects of curvature on surface tension. This creates a vapor pressure gradient that drives diffusion of vapor to the ice-crystal surfaces. Ice crystals grow at the expense of water droplets in something known as the Bergeron process. As long as sufficient moisture is available in a cloud, ice crystals—snowflakes—will continue to grow until they become heavy enough to drift to the ground. Temperatures in the lower troposphere dictate whether precipitation will fall as rain, snow, or something in between (sleet).

The phase of precipitation is not always easy to predict; snowfall is commonly recorded at near-surface temperatures well above 0°C, and rain can occur at temperatures below 0°C. This depends on the temperature structure in the atmospheric boundary layer, the initial size of the precipitation particle, and the transit time (height, speed, and path) of the precipitation. In climate downscaling applications where one needs to estimate snowfall from the daily or monthly precipitation and temperature, it is recommended to estimate the fraction of precipitation to fall as snow as a statistical distribution about 0°C, rather than adopting a sharp transition at, for example, 0°C or 2°C, below which precipitation falls as snow. For instance, a cumulative distribution function over the range of approximately –6°C to +6°C represents observations well. Ideally, of course, atmospheric models are able to represent explicitly the precipitation processes and phase of precipitation on the temporal and spatial scales needed for modeling snow accumulation.

The structural forms of snowflakes are determined by the temperature, humidity, and wind conditions during crystal growth, although the specific processes that determine a snow crystal's form remain somewhat enigmatic. The many different crystal forms include pillars, disks, stellars, plates, and columns. The relation to cloud conditions is complex and nonintuitive, but some systematic tendencies are observed, with useful

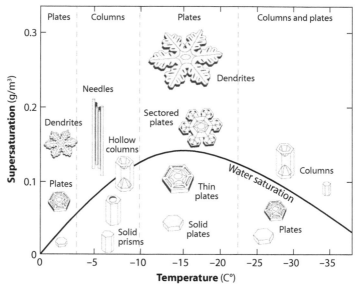

Figure 4.1. Snow crystal morphology as a function of cloud temperature and supersaturation. (Adapted from Y. Furakawa by Libbrecht (2003). Figure courtesy of K. Libbrecht.)

generalizations given in figure 4.1. Faceted hexagonal plates are prevalent under dry conditions below –10°C and again at temperatures above –4°C, but columnar ice crystals are dominant between –4 and –10°C. When humidity is higher (i.e., under supersaturated, relatively mild conditions), plates evolve into dendrites and stellar crystals. Ice needles develop around –5°C, and columns prevail under cold, humid conditions, below ca. –20°C.

Ice crystals in the cold, dry upper troposphere tend to be hexagonal plates. These crystals align horizontally due to air resistance, and the hexagonal structure can reflect sunlight at specific angles (22°) to produce halos or sundogs. They also give us elegant cirrus clouds. Closer to the ground, more water vapor is commonly available, and supersaturated clouds produce the classical dendritic forms.

Snow crystals inherit their hexagonal macrostructure from the structure of the ice lattice, and humidity plays the primary role in shaping the growth of elaborately branched, stellar crystals from six-sided, faceted plates. Small plates develop when snow-crystal growth is moisture-limited and vapor is deposited equally on all six facets. With surplus moisture, vapor deposition occurs rapidly and near-symmetrically at the corners of the facets (the shortest distance for vapor diffusion) and a branching instability is excited, where dendritic branches grow at each corner. New branches are spawned at asperities and protrusions in the branches. The snowflake grows until it becomes heavy enough to float to the ground.

Mechanisms of Snowfall

Snowfall is associated with two main mechanisms of precipitation: (i) orographic uplift of moist air masses and (ii) air mass mixing/frontal precipitation. The former is dominant along mountainous coastlines, where inland advection of maritime air masses leads to uplift, cooling, and a transition from rainfall at low altitudes to snowfall at higher elevations. Snow-capped peaks in New Zealand, Iceland, Norway, the Himalayas, and the American Cordillera all testify to this process. Orographic precipitation also delivers snowfall to the flanks of the Greenland and Antarctic ice sheets.

Frontal collisions produce much of the snowfall in interior continental regions, particularly in the mid-latitudes where extratropical cyclones usher a steady succession of cold, polar air masses into contact with lower-latitude (warmer, wetter) air masses. Mark Helprin expresses this snowfall mechanism sublimely in *Winter's Tale*:

> Battalions of arctic clouds droned down from the north to bomb the state with snow, to bleach it as white as young ivory, to mortar it with frost that would last from September to May.

Although violent, this metaphor is consistent with the meteorological appropriation of the term *front* from its military origins. In the case of winter snow storms, conflicting air masses produce precipitation in several ways. Dense polar air in cold fronts plows under warmer air masses, forcing uplift, cooling, condensation, and precipitation. Developed extratropical cyclones are also accompanied by eastward or poleward movement of warm fronts, with warm, wet air masses buoyantly overriding cool air at the surface, again inducing cooling and precipitation over a region. In some cases, forced uplift along cold or warm fronts triggers free convection and even stronger storms (low pressures, high winds, and large amounts of moist air advection to a region). From fall to spring, these frontal interactions produce snowfall as the precipitation falls through cold air near the ground, although other forms of precipitation are also common.

Of course, orographic and frontal precipitation processes frequently act in concert. A common example is when moist Pacific air masses (often associated with extratropical cyclones) are forced upward by mountainous topography in New Zealand or western North America. Other precipitation processes also give rise to snowfall, such as air mass modification (e.g., due to addition of moisture from open water bodies and/or isobaric cooling of air masses as they move poleward or inland).

Variations on Snowfall

There are many other forms of meteoric precipitation in the solid phase. *Sleet* is the name for liquid precipitation that has partially frozen during its descent through the lower atmosphere; it is amorphous, lacking the crystal structure of snowflakes. Freezing rain occurs when raindrops freeze upon contact with cold surfaces on the ground. *Rime* is an extreme form of this: supercooled water droplets that freeze on contact and can produce a thick, white coating over trees, power lines, or other objects that intercept humid, near-surface air flow. *Graupel* is a term for snowflakes that are partially melted and rounded into pellets and can be coated in rime. These are also called snow pellets, and you can recognize them because they often bounce upon impact. *Hailstones*, in contrast, do not begin life as snowflakes. Rather, hailstones are dense, amorphous pieces of solid ice, formed from riming and deposition onto ice nuclei in cumulonimbus clouds. Strong updrafts promote cycling of ice particles in these clouds. Large hailstones result from high humidity and long residence times and can sometimes be found with concentric growth layers.

Frost is a variation on solid precipitation. It is the solid-phase form of dew, associated with direct deposition of water vapor onto a surface when a near-surface air mass is cooled to saturation. The source of moisture in frost is usually atmospheric, but there are also splendid examples of *surface hoar* deposits or *frost feathers*, in which the vapor pressure gradient between a warm, wet surface and a cold, dry air mass drives vapor diffusion from the surface to the air. Upon contact with the cold air, this vapor freezes to form intricate frost patterns that rival those of snowflakes.

Snow on the Ground

Although there are several different types of solid precipitation, there are even more variations and appellations for snow on the ground. Mariana Gosnell discusses the veracity of the legendary 100-plus different words for snow in Inuktitut, confirming only about 20 of these in conversation with Inuit elders. Dictionaries of more than 120 variations can be found, inflated by the conjunction of nouns, adjectives, verbs, and adverbs into single words describing snow and ice. Table 4.1 provides some examples. Regardless of the semantic debate, the Inuit people know whereof they speak; they read the story of the weather, seasons,

Table 4.1
A Sampling of Inuit Terms for Snow and Ice

Inuktitut	Meaning
Alutsiniq	Deep snow hollow
Aqilluqqaq	Soft snow underlying crust
Atairranaqtuq	Squeaky snow
Aujaqsuittuq	Eternal snow
Kaiyuglak	Rippled surface of snow
Masak	Slush
Mauya	Snow that can be broken through
Munnguqtuq	Compressed snow softening in spring
Mitailak	Soft snow covering an opening in an ice floe
Niummak	Deep, soft, newly fallen snow
Perksertok	Drifting snow
Pukaq	Uniformly soft snow
Pukak	Sugar snow
Pukajaaq	Granular/crystallized snow
Qanniq	Falling snow
Qimugjuk	Snow drift/shaped snow
Qingainnguq	Brilliant ice crystals falling
Qiqirrituq	Snow squeaky once
Qiqumaaq	Snow with frozen surface
Qiqsuqaq	Glazed snow in thaw time
Quna	Slush ice
Siku	Ice
Sikuliaq	Youngest ice
Sikussaq	Thick, deformed, multiyear sea ice that forms in fjords
Sikuuttuq	Freshwater ice

landscape, and hazards in the subtle variations of the snow and ice, and the rhythm of traditional Inuit life is tuned and adapted to the cryosphere's mood.

Explicit in the Inuktitut lexicon are descriptors for the density, viscosity, age, hardness, and water content of snow and the spatial variability of snow on the ground: uniform, hummocky, drifted, crusted. These features also shape the interactions of snow within the climate system. The density and depth of fresh snow determine its thermal (insulating) properties, which are important to growth of sea ice, freshwater ice, and permafrost. Snow water content is the essential parameter of interest for glacier mass balance and water resources applications (snow hydrology). The seasonality of snowfall and the locations where snow preferentially accumulates on the landscape influence all of these systems. These also impact society through avalanche hazards, transportation disruptions, and recreational opportunities (i.e., ski areas).

As discussed in chapter 2, initial snow density is variable and is a function of the moisture content, temperature, and wind conditions during snowfall; 100 kg m^{-3} is typical. Once snow is on the ground, these same meteorological variables play a strong role in grain-scale snow metamorphism and densification of the snowpack. In cold, dry conditions, sustained winds of several meters per second will harden or stiffen fresh snow into something colloquially called "wind slab," which reaches densities of 400 kg m^{-3}. Wind slab has rounded, well-packed snow grains, with uniform and "fine" grain sizes of less than 0.5 mm. This snow can generally support a person or sled. In extreme cases, strong winds and a dearth of fresh snow can lead to continued evolution of surface snow into a sculpted, extremely hard version of wind slab known as *sastrugi*.

In the absence of strong winds, cold snowpacks can remain relatively light for months. With time and with buildup of the snowpack, grain sintering in buried snow brings about gradual densification. This is the slow rounding of crystals and development of intergranular bonds through systematic vapor transport from angular to rounded crystals. These processes are accelerated at warm temperatures. Weakly bonded stellar and angular crystals can persist throughout the winter, until thawing of the snowpack introduces meltwater and rapid grain metamorphism.

Vapor diffusion and deposition within a snowpack also leads to postdepositional modification of the snow stratigraphy. Moisture from the atmosphere, the underlying ground, or from sublimation or evaporation within the snowpack itself moves upward or downward along pressure gradients, from warm to cold regions of the snowpack. This vapor can be deposited on contact with the colder snow grains. In most seasonal snowpacks, this is an upward movement from geothermally warmed lower layers in the snowpack to the atmospherically cooled layers above. Because of snow's low thermal conductivity, it can support strong temperature gradients (e.g., $>10°C\ m^{-1}$), which are conducive to strong vapor pressure gradients and vapor transport. This produces *depth hoar* in a snowpack: the growth of large, faceted, and poorly bonded snow grains.

Depth hoar that forms in Arctic snowpacks in the fall and early winter is often a few decimeters deep, and the grains can remain faceted, with sizes of a few millimeters, until the following summer, giving a "hollow" layer beneath the overlying winter wind slab. In lower-latitude mountain snowpacks, depth hoar creates weak layers that can also persist for several months and are implicated in deep slab avalanches.

Wind Redistribution

As well as acting as an agent of densification, wind mobilizes and transports snow. Emerson's *The Snowstorm* beautifully captures the "frolic architecture" of snow that is shaped by winds.

Blowing snow gives blizzards their distinct, disruptive character, with horizontal advection of snow particles obscuring visibility and leading to what can seem like several hours of snowfall with negligible snow accumulation on the ground. The stronger the wind, the more loose snow that can be mobilized. Deposition is a similar process to *saltation* of dust particles or other aerosols transported in the atmosphere. Blowing snow creates snowdrifts in places where wind flux is convergent or where turbulent eddies result in downward advection and snow deposition. The classical example is snow fences, which create eddies that induce snow accumulation downstream of the obstruction. Natural topographic variability produces the same effect; snow is scoured from the upwind side of obstacles and deposited in the lee side or in hollows and cavities. This has the effect of smoothing out the topography. In mountainous terrain, snow is similarly scoured from the upstream side of ridge tops and deposited in treacherous *cornices* (overhanging snowdrifts on the ridge top) and on the lee slopes that lie below.

Wind-blown snow is mainly redistribution over an area for a given snowfall event. Over a large enough area, the depth and water-equivalence of snow on the ground averages out and is representative of the amount of precipitation that fell from the atmosphere. This can be important to glacier mass balance or water resources studies, which are based on the total snow on the ground within a given catchment; redistribution affects local snow depth and the timing of melt, but may not have a strong effect on the total annual mass/water budget. However, some blowing snow may be carried large distances by the wind and redeposited off-glacier or in a different hydrological catchment, removing it from the system. Similarly, some wind-blown snow is subject to blowing-snow sublimation, a process that can remove mass from the system.

Snow on the ground is also lost to direct sublimation into the atmosphere. This is a function of wind speed, humidity, and boundary-layer turbulence, which promotes exchanges of momentum, mass, and energy between the surface and the atmosphere. As introduced in chapter 3, the surface (skin) layer over snow and ice is typically assumed to be saturated, with the saturation vapor pressure determined by the snow/ice surface temperature. If vapor pressure in the overlying air is less than this, the vapor pressure gradient supports sublimation (or evaporation, if liquid water is present). Condensation or deposition from the atmosphere occurs in the opposite situation, and when it is cold and calm this leads to the formation of elegant hoar frost crystals in the surface snow, as described earlier.

Observations of Seasonal Snow

Detailed information about snowfall is available from direct measurements in populated regions of Europe, Russia, and North America over the past century. Much of this information is archived as snow depth or fresh snowfall totals, with snow–water equivalence (SWE) estimated based on the assumption that fresh snow has a density of 100 kg m^{-3}. The most reliable methods of measuring SWE are to melt the collected snowfall and measure it as if it were liquid precipitation or to sample it directly for snow density (i.e. through weighing of snow samples of known volume). Such methods became routine in the middle of the 20th century, along with the introduction of regular snow surveys in many of the world's mountain regions.

Beginning in the 1960s, visible-wavelength satellite imagery has provided exceptional mapping of snow-covered area in the Northern Hemisphere. Available station data allow Northern Hemisphere snowpack trends to be extended back to 1922, though with less confidence. For the period 1967–2010, the satellite era, the mean Northern Hemisphere snow-covered area was 24.9×10^6 km^2, peaking in January (46.7×10^6 km^2) and reaching an annual minimum in August (3.1×10^6 km^2).

Most of the residual summer snow in the Northern Hemisphere is on the Greenland ice sheet. Additional snow falls on the oceans at high latitudes but is quickly melted, excepting that which accumulates on sea ice. Snow on sea ice is not accounted for in the snow cover data.

Figure 1.3 plots the average winter snow cover in each hemisphere. This geography closely follows the 0°C isotherm on the landscape. Given cold enough temperatures for snow, the thickness and SWE of seasonal snowpacks are generally proportional to the available moisture in a region. This is a function of the temperature of the air, continentality (distance from the ocean or other major water bodies), and the prevailing winds.

Maritime environments at midlatitudes see large accumulations of snow, as these settings offer the ideal combination of cool winter weather, ample moisture supply, and baroclinic eddies that advect this moisture inland. Precipitation rains out as air masses cool over the continents. Latitudes of the polar front (the polar jet stream) are optimal for this, as meridional meandering of the jet stream generates repeated cyclonic disturbances along winter storm tracks. As discussed earlier, this produces heavy snowpacks where these weather systems intersect elevated topography (e.g. the coastal mountains of North America, Norway, Iceland, and the Southern Alps of New Zealand). Annual snow accumulations in excess of 2000 mm w.e. are common here. Comparable snow accumulation totals are possible on the windward side of other major orographic features like the southeast margin of the Greenland ice sheet and the southern slopes of the Himalayas.

At tropical latitudes there are high quantities of convective precipitation, and orographic uplift is also active in many regions, but temperatures are generally too warm to permit snowfall. Only the highest elevations (e.g., above 5000 m) receive significant snow accumulation. In contrast, air temperatures are deeply cold at high latitudes of the Arctic and Antarctic, below –20°C for most of the year, and this limits the available moisture content of the atmosphere. The Clausius–Clapeyron law introduced in chapter 2 is an expression of this thermodynamic relationship. Writing this in terms of vapor pressure,

$$\frac{\partial e_s}{\partial T} = \frac{L_v}{T_a \Delta a_v}, \tag{4.1}$$

where e_s is the saturation vapor pressure, T_a is the air temperature, L_v is the latent heat release of the phase change from vapor to either water or ice, and Δa_v is the specific volume change during the phase change. Equation (4.1) is more commonly expressed in its integrated form, where it provides a relation describing saturation vapor pressure as a function of temperature. Because water vapor occupies such a high volume relative to liquid water or ice, $\Delta a_v \approx a_v$, and the ideal gas law can be invoked to allow expansion of (4.1). Several approximations exist, with varying degrees of accuracy, but for most applications the World Meteorological Organization recommends the following forms:

....

$$e_{sw}(T_a) = 6.112 \exp\left(\frac{17.62T_a}{T_a + 243.12}\right), \tag{4.2}$$

$$e_{si}(T_a) = 6.112 \exp\left(\frac{22.46T_a}{T_a + 272.62}\right), \tag{4.3}$$

where T_a is in degrees Celsius, and the vapor pressures have the unit millibar. The derivation of these expressions can be found in most texts on atmospheric thermodynamics. More elaborate parameterizations are available, but Eqs. (4.2) and (4.3) are accurate to within 1% for most Earth-surface temperatures. Higher-order approximations are recommended at very cold temperatures (e.g., below −40°C).

The strong sensitivity of saturation vapor pressure to temperature provides a hard limit on precipitation in polar regions. Cold air is simply too dry to provide large amounts of snowfall. The interior regions of the Antarctic and Greenland ice sheets are polar deserts, with annual precipitation totals of less than 250 mm w.e. Continental polar tundra environments in Canada and Russia are similar; for instance, Eureka Nunavut (79°59′ N, 85°56′ W) had an average annual precipitation of 70 mm w.e. from 1950 to 2009, with 52 mm w.e. falling as snow.

Similar aridification is found in the lee of mountain regions at lower latitudes. For example, annual snow accumulation in the Chilean Andes, on the eastern side of the Rocky Mountains, or on the northern slopes of the Himalayas is about an order of magnitude less than that on the windward side. Hydrometeors advect and drift with the wind, with downstream advection of tens of kilometers. This augments the snowpack in the vicinity of continental divides and in the upper regions of lee slopes. Aridity sets in downstream of this. This depends on the moisture content and temperature of an air mass; large quantities of moisture may survive the passage across low orographic obstacles or the initial ranges of mountain chains.

For orographic barriers that are not too large, snow accumulation usually increases with elevation on windward slopes, with steep annual snow accumulation gradients: 1000 mm w.e. km^{-1} is typical, though this is highly variable and is ultimately limited by the available moisture. For especially high mountain barriers, such as the Andes, the Himalayas, or the Saint Elias Mountains, drying of the air mass occurs at high elevations on the windward side. In this case, precipitation and snow accumulation usually increase with elevation to a point (e.g., 3000 to 4000 m), then decline above this altitude. These precipitation processes and snowpack gradients are a steep challenge for climate models, which are unable to resolve mountain topography in sufficient detail.

Snow on the ground in mountain environments is also subject to avalanching and a strong degree of wind redistribution. This tends to limit snow accumulation on steep slopes and ridge tops while promoting snow deposition on the gentler slopes of alpine valleys, meadows, and forests. These environments naturally give rise to mountain glaciers.

Proxies for Seasonal Snow Cover

Because snow extends to elevations and latitudes where it is near the threshold for viability—temperatures close to 0°C—snow cover is very sensitive to warming or cooling. The record of recent changes in snow-covered area is discussed in chapter 9. There is scattered observational data on snowpack variations prior to the 20th century—anecdotal or through direct observations at select sites (typically measurements of snow depth). There is no direct geological proxy for past snow cover, although lichen-free zones in tundra environments provide indications of areas that were covered by snow or ice for extended periods prior to the 20th century. The timing and extent of spring runoff in seasonally snow-covered catchments can also leave distinctive traces in lacustrine sediments. Certain tree species are sensitive to spring snowpack through its influence on soil moisture and the onset of the growing season; hence, some tree-ring records offer indirect reconstructions of past snowpack variability.

Ice cores from glaciers provide direct records of net snow accumulation that can be annually resolved. These records extend back several centuries or millennia in mountain icefields or hundreds of thousands of years in the polar ice sheets. In the latter case, there is negligible summer melt so net accumulation is equivalent to the annual snow accumulation. In melt-affected environments, like most mountain regions, ice cores record the net annual snowfall less the amount of meltwater runoff:

hence, a combination of precipitation and temperature influences. Where it is possible to isolate a precipitation signal, it can be difficult to interpret this on a regional scale. For most of the Earth's surface, precipitation is a mixture of snowfall and rainfall. Warmer and wetter conditions in a region may register as an increase in snow accumulation at high altitudes but a loss of snow at low altitudes, as rainfall increases and snowfall declines.

At the highest elevations and latitudes, precipitation roughly equates to snow accumulation, and ice cores acquired from these latitudes may offer regional- or even synoptic-scale insights into precipitation variability. For instance, snow accumulation records from ice cores in the Saint Elias Mountains in Yukon and Alaska are heavily influenced by low-frequency variability in meridional circulation in the north Pacific, which is affected by El Niño–Southern Oscillation (ENSO) and decadal- to century-scale variability in Pacific sea surface temperatures. Snow accumulation records from these sites therefore include the influences of climate variability on a spectrum of timescales, which needs to be decoded.

SNOW-MELT MODELS

Models of snow melt are needed for regional-scale hydrological applications such as hydroelectric power generation, flood hazard, and water resources forecasts. They are also necessary for modeling of cryospheric interactions with the climate system and global cryospheric response to climate change.

A solution of the full surface energy balance described in chapter 3 is desirable for predicting rates of snow melt, but this requires extensive meteorological data that are generally unavailable for the remote regions and spatial scales of interest. For this reason, simplified snow-melt models are commonplace, both operationally and in glaciological and climate models. Positive degree day ("temperature index") melt models are the most simple and popular tool. These work on the premise that snow melt is linearly proportional to air temperature when $T_a > 0°C$, such that the total melt over a period of time τ scales with the cumulative positive degree day total over this time:

$$m(\tau) = b \cdot \text{PDD}(\tau) = b \int_0^\tau H(T_a) T_a(t) \, dt, \qquad (4.4)$$

where PDD is the positive degree day total over time τ (measured in units °C · d), b is an empirically determined melt factor, and $H(T_a)$ is a Heaviside function, equal to 1 when $T_a \geq 0°C$ and equal to 0 when $T_a < 0°C$.

This relationship works surprisingly well in estimating monthly or seasonal snow and ice melt, as long as b is appropriately tuned for a site. The explanation for this is that the main sources of energy that drive melting—solar radiation, incoming longwave radiation, and sensible heat flux—are all strongly correlated with near-surface air temperature. Longwave radiation and sensible heat are proportional to air temperature, and diurnal cycles of air temperature are broadly driven by shortwave radiation.

Degree-day melt models avoid the intensive meteorological data that are needed for energy balance calculations, requiring only air temperature as an input. Air temperature is believed to be relatively simple to extrapolate from remote point data or large-scale model data, using atmospheric lapse rates to account for elevation effects. However, both melt factors and lapse rates vary spatially and temporally, and degree-day models are far removed from the physics that describes the melt process. They can perform poorly when not locally calibrated or when applied to short timescales (e.g., hourly or daily melt).

Radiation-temperature index melt models offer an increasingly popular alternative that captures more aspects of the actual energy balance. These models take advantage of the fact that potential direct solar radiation can be calculated over the landscape, following Eq. (3.9), such that this likely influence on melt energy can be added to degree-day models with no additional meteorological data demands. The adjustment to (4.4) takes the form

$$m(\tau) = \left(aQ_{s\phi} + b\right) \cdot \text{PDD}(\tau), \tag{4.5}$$

where a and b are empirical parameters. An obvious weakness of (4.5) is that it does not account for the true absorbed solar radiation at the snow surface, which can deviate a great deal from the potential direct radiation as a result of cloud cover and fluctuations in surface albedo. These effects are embedded in the parameter a, and (4.5) can be rendered more physically based by explicitly building in temporal and spatial variations in surface albedo. It is also possible to derive regional cloud cover indices from satellite imagery or climate model diagnostics, allowing the first term in (4.5) to be replaced by an estimate of the absorbed radiation, $Q_S^{\downarrow}(1 - \alpha_s)$, which directly governs snow melt.

Alternative simplifications to the full surface energy balance have also been adopted, and improvements in mesoscale climate models as well as satellite observing systems are creating the possibility for a more physics-based approach in regional and global-scale snow-melt models. However, there is a perpetual trade-off between physics-based approaches, which use artificial, potentially flawed meteorological input data, and empirical models, which have more reliable input data but are missing some important physical processes. Empirical models are not very portable in space or time, so they are dubious for high-resolution distributed models and for future projections; the research community is therefore driving toward a more complete and faithful representation of the surface energy balance.

Water Flow in Snowpacks

Meltwater remains part of the snow or ice system until it runs off. Meltwater can refreeze in a snowpack or it can be stored for days to weeks in the liquid phase, delaying runoff to river systems. The seasonal snowpack is sometimes called a "snow aquifer" for this reason. A small amount of liquid water content increases the viscosity and cohesion of the snow, giving "packing snow" that is ideal for backyard snowmen. Snowpacks retain liquid water contents of several percent in the pore space before they become saturated and turn to slush.

Once saturated, gravity will triumph over surface and capillary tension, allowing water to drain through from the snowpack. Drainage is essentially Darcian: flow in a porous medium, driven by gravity and hydraulic pressure gradients. Percolation and drainage also occur in unsaturated snowpacks where vertical drainage channels, or "pipes," establish.

Water that drains to the base of the snowpack can pond there, infiltrate into the soil and groundwater system, or run off as a surface (overland) flow. If the underlying ground is frozen or in the case of a supraglacial snowpack, the snow rests on an effectively impermeable barrier, and runoff occurs at the snow–ground or snow–ice interface. This is common in the early melt season; one hears the water running beneath a tenuous snow cover, and creek crossings on such a snow cover should be attempted with care. If there is a topographic slope (hydraulic gradient), meltwater can also drain as a surface or internal flow in a saturated snowpack. Most of the meltwater in subsurface and surface flows drains to regional river systems. This is relatively quick for overland flows (hours to weeks), but there can be delays of days, months, or even years for water that enters the groundwater system.

Much more could be written about seasonal snow, and it will not stray far from our minds as we continue on with overviews of freshwater and sea ice, glaciers, and permafrost. All are heavily influenced by the seasonal snowpack; glaciers would not be here without snow, and snow insulates the other forms of ice from the atmosphere, playing a critical role in their growth, decay, and thickness. Heavy snow cover on a surface of lake or sea ice can even result in submergence of the ice cover, with flooding and subsequent freezing of the snow creating a coating with the doubly cryospheric name "snow ice." The rest of this chapter expands on this and other important aspects of lake and river ice.

RIVER AND LAKE ICE

At the same time as seasonal snow spreads from the polar regions to the midlatitudes each winter, a veil of seasonal ice forms over rivers and lakes. The essential thermodynamics of this process are described in chapter 3. This section expands on this through a summary of several of the main features of lake and river ice.

The ice season for rivers and lakes is defined by *freeze-up*, the moment of the year when continuous ice cover sets in, and *breakup*, the time when seasonal ice wastes to the point that open water becomes extensive. In rivers, ice advection is often factored into the definition; freeze-up may be defined as the time when ice cover becomes immobile, whereas breakup is heralded by the time when ice begins to move downstream. These definitions clearly contain some ambiguity. For instance, "extensive" open water may refer to the first occurrence of visible open water, to the time when travel becomes hazardous on the ice, or to the time when ship navigation becomes possible.

Lake Ice

Lake ice forms due to cooling and freezing from above, driven by radiative and sensible heat losses to the atmosphere in the autumn and winter. The surface energy balance described in chapter 3 is applicable here. Snow falling in open water can provide an additional energy sink, due to the latent energy removed from the lake to melt the snow.

If temperatures reach 4°C as a lake cools, the density maximum of water at this temperature (see figure 2.2) results in curious behavior. Dense surface waters will sink, creating a systematic fall cycle of "flushing." Overturning in the water column brings warmer water to the surface, where it will cool as the flushing process continues. There can be several flushing events in a lake each autumn, until much of the water column has a temperature near 4°C. Continued cooling creates stably stratified surface waters that will begin to freeze. Once ice is nucleated, conductive heat loss to the atmosphere promotes ice growth. Growth rates are sensitive to the depth (insulating capacity) of local snow cover and the temperature gradient, hence the air temperature. This is illustrated below.

Ice cover is intermittent in regions where winter temperatures fluctuate about 0°C. Cold spells (sustained temperatures of less than 0°C over a period of several days) set up a tenuous layer of ice that is millimeters to centimeters thick. Thin ice cover has little thermal inertia, so a brief (e.g., 1- or 2-day) warming episode can melt the ice. Such freeze–thaw cycles characterize many lakes all winter long.

If temperatures remain subzero for long enough to form a thicker ice cover, however, it requires a large amount of latent heat energy to melt the ice. Winter nights are long at high latitudes and are characterized by a net energy deficit (primarily longwave cooling), so ice thickens overnight, and there are limited daylight hours to drive melting in midwinter. This makes a well-established lake ice cover quite endurable, such that it is able to weather multiday warm spells, and it will typically persist through to spring.

Freezing of lake water creates *congelation ice*, which is commonly called *black ice*. This is not to be confused with the "black ice" that forms on roadways when refrozen water forms a thin, transparent veneer of ice, making it treacherous for driving. Black ice in lakes is characterized by ordered, vertically aligned ice columns, a result of slow crystal growth driven by vertical heat diffusion (conductive heat losses upward into the overlying ice). Black ice is an ironic name; it has relatively few air bubbles, making it highly transparent to light. Hence, it is possible to see right through this ice into the underlying water, which is dark, giving the ice its black appearance. In contrast, ice that forms from above due to flooding of snow-covered ice is called slush ice, snow ice, or white ice. Its high bubble content promotes strong backscatter of light and an opaque white color. White ice has small, randomly ordered ice crystals.

As noted in the earlier section "Snow-Melt Models," flooding can occur if snow on the ice is thick enough to cause ice to sink below the waterline. Snow-ice can also form from saturation and freezing of snow cover due to lake water that is pushed up through fissures or *cryoconite holes* in the lake ice cover. Such fissures may be a result of contraction cracks that form on cold nights. Flooding can also be caused by groundwater flows or streams that feed a lake, where such water flows are dammed or diverted to flow on the surface of the ice cover. Icing from this effect is also called *naled* or *aufeis*, literally "ice on top."

Ice does not completely cover lakes in many cases, particularly in the early stages of freeze-up, as wave action and the motion of water in inlet or outlet streams prevent ice from setting up. Because of the greater wave energy in the middle of a large water body, lakes often freeze over from the edges, radially inwards.

Lake ice cover is surprisingly strong in its ability to support a load. A consistent 5- to 10-cm ice cover is enough to support a person on foot (e.g., skating or ice fishing), 12–15 cm permits snowmobile travel, 20 cm can support a light automobile, and more than 50 cm enables heavy truck transport. These are general rules of thumb, and things such as underlying water currents or fractures in the ice can compromise its strength. Different jurisdictions provide different guidelines for winter ice travel. In Finland, for instance, ice roads used for light vehicle transport must have an ice thickness of at least 40 cm across the entire expanse of the road before they are opened for the season.

Thermodynamics of Ice Growth and Decay

Ice growth proceeds quickly during early stages of freeze-up but is then self-limiting due to insulation of the basal ice–water interface from cold atmospheric temperatures. In a single growth season, this thermodynamic limitation gives maximum ice thicknesses of 1.5 to 2 m in extremely cold environments (i.e., polar regions). Snow cover has a strong additional insulating effect, further limiting ice thickness. Where seasonal ice roads are important for transportation, engineering is often done to stimulate thicker ice artificially. The two most effective tactics for this are to clear the snow cover and to spread water on the ice, introducing ice accretion from above. Similar techniques are used in building and maintaining backyard skating rinks.

The thermodynamics of lake ice follow the equations outlined in chapter 3, with the surface energy balance dictating the energy deficit or surplus that is available to drive ice growth and decay, and Eq. (3.5) governing the internal energy and temperature evolution in the ice. During ice growth, vertical heat conduction governs heat loss from the lake and ice into the atmosphere, subject to atmospheric temperature forcing at the ice/snow–atmosphere interface and heat flux from the lake into the ice at the ice–water interface. Basal ice accretion/ablation can then be modeled from the net surplus or deficit of energy at the base. For ice thickness H, this follows

$$-\rho_i L_f \frac{\partial H}{\partial t} = Q_{in}\big|_{b-} - Q_{out}\big|_{b+} = Q_w + k_i \frac{\partial T}{\partial z}\bigg|_{b+}, \qquad (4.6)$$

where L_f is the latent heat of fusion, and Q_{in} and Q_{out} refer to the upward-directed vertical heat flux into and away from the lake–ice interface (the base of the ice). Q_{in} is

93

equal to the heat flux from the water, Q_w, and Q_{out} is the heat conducted upward into the ice. The symbols $b-$ and $b+$ denote the infinitesimal layer below and above the base of the ice. The upward-directed heat flux is a function of the air temperature and the thickness and thermal conductivity of the overlying ice and snow.

When combined with Eq. (3.5) and using a moving (adaptive) grid in a numerical scheme that tracks the phase front, the model introduced in chapter 3 can be applied to simulate ice growth subject to air temperature forcing. Figure 4.2 plots an example for an idealized case with $Q_w = 0.5$ W m^{-2} and meteorological forcing based on observed air temperatures and snow depths in Yellowknife, northern Canada (62.46° N, 114.44° W). The solid line in figure 4.2c is for a hypothetical case with pure ice (no snow cover), and the dashed line includes a variable snow cover based on the Yellowknife meteorological data, assuming a snow density of 300 kg m^{-3}. I use temperature- and density-dependent values for the heat capacity and thermal conductivity of snow and ice as given in chapter 3.

Yellowknife is a cold, dry site that is representative of a subpolar, continental climate: ideal conditions for seasonal lake and river ice. Indeed, there is an extensive network of winter ice roads in this region. Figure 4.2 illustrates the limitations on growth imposed by both the ice thickness itself and an overlying snow cover. The highest rates of growth are early in the freeze-up season, when the ice and snow cover are thin. Even 10–20 cm of snow is enough to have a strong insulating effect on the

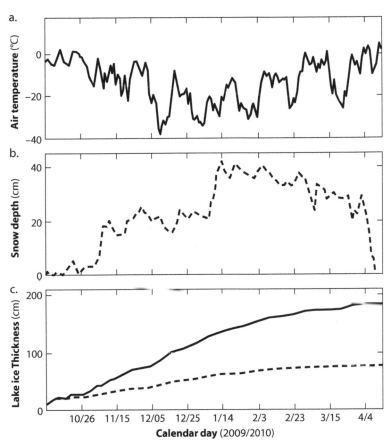

Figure 4.2. Modeled lake ice growth through a winter season, based on daily air temperatures and snow cover in Yellowknife, Northwest Territory, Canada, from winter 2009–2010. (a) Air temperature and (b) snow cover forcing used for this scenario, October 7, 2009, to April 25, 2010. (c) Lake ice thickness for snow-free conditions (solid line) and including the effects of the snow cover (dashed line).

ice, especially if it remains cold and dry; water-saturated snow is more conductive. The effective insulation is also very sensitive to snow density.

The growth of ice can be approximated via Stefan's solution, which is an analytical approximation to (4.6). Neglecting the heat flux from the lake, the effects of overlying snow cover, and assuming a linear temperature gradient in the ice,

$$H(t) = \sqrt{\frac{2k_i F(t)}{\rho_i L_f}}. \tag{4.7}$$

Here, $F(t)$ represents cumulative freezing temperatures, a concept similar to positive degree days in Eq. (4.4) but measuring the integrated temperature below 0°C. In (4.7), this has units °C · s, though freezing degree days (negative degree days) can be used instead, with a conversion in the units for thermal conductivity.

Figure 4.3 plots the solution to Stefan's equation for an idealized scenario: a step temperature change to −10°C, maintained for 200 days. For comparison, the numerical solution is also shown here, as calculated from (4.4). An additional curve is added to indicate the additional impact of snow cover on ice growth, assuming a linearly increasing snowpack over the 200-day period of ice growth and a maximum snowpack thickness of 40 cm.

This analytical solution has also been used in models of sea ice and permafrost growth. It indicates that ice growth follows a square root of time function, assuming that $F(t) \propto t$ during the winter season. By corollary, $dH/dt \propto t^{-1/2}$. In practice, where (4.7) has been used operationally

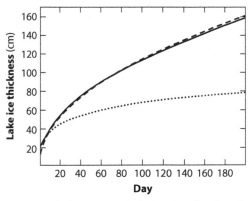

Figure 4.3. Lake ice growth during 200 days based on the Stefan solution (dashed line) versus the numerical solution from Eq. (4.4) (solid line), for a step cooling to an air temperature of –10°C. Lake ice thickness is also plotted for a case with linearly increasing snow cover over the 200-day integration, to a maximum snow depth of 40 cm (dotted line). A snow density of 300 kg m^{-3} is assumed.

for lake ice, a coefficient f (ranging from 0.2 to 0.8) has been added to the right-hand side of (4.7) to reduce the predicted ice depth in order to account empirically for the insulating effects of snow and other assumptions inherent in (4.7). Higher numbers apply in windy environments with little snow cover, whereas low values apply where snow cover is deep and subsurface heat fluxes are high (e.g., with warm underlying lake waters).

An alternative simplification of Eq. (4.6) introduces a two-layer model to approximate the effects of snow cover. For ice thickness H_i and snow depth d_s,

$$\rho_i L_f \frac{dH_i}{dt} = \frac{F(t)}{H_i/k_i + d_s/k_s}.$$ (4.8)

This is essentially an extension of Stefan's equation, treating layers of ice and snow in series with respect to heat conduction. This approximation assumes linear temperature gradients in the ice and the snow, with an equal conductive heat flux through each medium such that $\partial T_i/\partial z$ and $\partial T_s/\partial z$ have equilibrium values in inverse proportion to the thermal conductivities, k_i and k_s. Equation (4.8) can be solved to give the incremental ice growth, δH_i, over a time period δt with negative degree days $F(\delta t)$ [where $F(\delta t) = -\int T_a \, dt$]. It is also possible to add a third "resistor" to the denominator in (4.8), describing the effective heat transfer between the snow surface and the atmosphere.

In fact, there is little need to rely on Stefan's equation or (4.8) nowadays, as a full numerical solution of one-dimensional heat diffusion in Eqs. (3.5) and (4.6) is straightforward. For instance, the computer program to generate the examples presented in chapters 3 and 4 took less than a day to develop. Data concerning snow depth and the surface energy balance terms (Eq. 3.1) are much more limiting in simulation of lake ice.

Ice Decay

The thermodynamics discussed above also apply to basal melting of lake and river ice, but heat fluxes from the water do not usually drive this process. Rather, ice breakup—the spring thaw—is driven by the intensification of solar radiation in the spring and early summer, supplemented by increased longwave and sensible heat supply as the air and the surrounding terrain warm. The surface energy balance equations of chapter 3 are salient here.

Initial melting of freshwater ice proceeds slowly, accelerating in late stages due to the presence of low-albedo melt ponds and areas of open water. At higher latitudes, where the ice cover is thick, lake ice tends to persist longer into the spring than the regional snow cover. Late in the melt season, warm air advection from adjacent snow-free land thus provides an additional source of heat energy to accelerate ice breakup.

It is common to find a layer of meltwater overlying lake ice that is still quite solid; the ice cover remains impermeable until late in the spring thaw. During advanced stages of ablation, internal melting from absorbed solar radiation transforms the ice surface to a pocked, lower-density medium that is colloquially referred to as "rotten" ice. Deteriorated surface ice of this nature can also be found on sea ice and glaciers in summer. Another feature that is often seen as lake ice decays is the development of *candle ice*, where columnar black ice is prevalent. When lake water freezes, impurities are rejected and collect at the interstitial grain boundaries. These impurities depress the freezing point at the junctions between vertically oriented ice crystals, and during initial stages of thaw these intercrystalline boundaries are the first regions to melt, isolating the columnar ice candles. Similar features can be seen in terrestrial icings that are formed from groundwater seepage through the winter.

River Ice

The seasonal cycle of river ice growth and decay has much in common with that of lake ice, but there are some important differences in the thermodynamics and the character of the ice. The flow of water creates some complexities. Because rivers are shallow and well mixed, the water throughout a reach of a river must cool to 0°C before it will begin to freeze (unlike lakes, in which freezing begins when much of the water column is at 4°C). Turbulent heat dissipation and conversion of potential energy to kinetic energy as a river flows downhill provide internal sources of energy that help river water to resist freezing. It is therefore common to see ice along the edges of a river and a coating of frozen spray on vegetation surrounding the river while the main channel is still flowing freely well into the winter.

Once nucleated, ice floes on a river advect downstream, so there is generally a mixture of open water and ice cover during early stages of freeze-up. Ice convergence in reaches with slower flows generates a more complete ice cover, and when conditions are cold enough the ice cover becomes fixed in place, anchored by *fast ice* that is rooted to the shore. At this stage, the evolution to fixed and complete ice-covered conditions often propagates upstream, and the ice can then thicken in place in similar fashion to the growth of lake ice. Water continues to flow beneath the ice in most major rivers, but some alpine streams and high-latitude rivers will freeze to the bed.

The complex channel geometry of rivers makes for large spatial and temporal gradients in ice thickness and compactness, particularly where ice cover is discontinuous. Ice can raft and ridge in the river channel or it can override the banks. River ice can also pile up, ground, and block the downstream flow, leading to ice-jam floods. Such floods can be sudden and extensive, overriding the high-water levels of "normal" spring freshet floods in high-latitude rivers. Ice-jam floods often occur where rivers widen and shallow, although they are notoriously difficult to forecast. River ice is unpredictable and hazardous during breakup, with shifting and overturning floes and frequent "surges" of ice from upstream.

River ice has other capricious characteristics, particularly where fast-flowing water suppresses freezing. Water can become supercooled in this situation, giving rise to the production of *frazil*—thin, hollow ice needles that can be found throughout the water column and are prone to conglomeration on any subzero surface that provides a good nucleus for ice. Rapids and open-water reaches of large streams and rivers can produce prolific amounts of frazil in a winter. Where this ice nucleates or agglomerates on subsurface rocks, the river bed, or built structures, it is called *anchor ice*. This often takes the form of flat, congealed platelets, but it can build up to large masses of ice (e.g., meters in diameter).

Observations of Freeze-Up and Breakup

River and lake ice have been monitored for many decades. The ice season influences transportation, recreation, weather, hunting, and other aspects of society in midlatitude and high-latitude communities. For these reasons, there is a detailed historical record of ice freeze-up and breakup in many communities, particularly where ship navigation and ice roads are dependent on ice conditions. In recent decades, there are also comprehensive satellite-based observations of lake ice cover and the ice season. Both visible and microwave imagery offer relatively straightforward discrimination between open water and ice conditions, as well as detection of meltwater ponds. Resolution limitations as well as high-frequency variability (advection of ice floes) make monitoring of river ice more difficult. Observations of ice thickness are also difficult and sparse. Chapter 9 describes the observational record of freshwater ice freeze-up and breakup for recent decades.

SUMMARY

This chapter has provided an introduction to some of the main features of seasonal snow and ice on the land. Building on the thermodynamics of the cryosphere introduced in chapter 3, the equations for the growth of lake and river ice have been presented and applied in simple examples. These equations are also applicable to the growth and thickness of sea ice and permafrost, so we will continue to build from this in the coming chapters.

The seasonal blanket of snow and ice unfurls over a large fraction of the world's land mass each autumn and retreats each spring, a cycle that affects weather, climate, ecology, and society. Seasonal snow is the most far-reaching but most ephemeral aspect of the global cryosphere. Its influence in midlatitudes gives it a powerful role in cryosphere–climate interactions. We return to this in chapter 8, and chapter 9 examines recent trends in terrestrial snow and ice cover.

> Land properly speaking no longer exists,
> nor sea nor air, but a mixture of these things,
> like a marine lung, in which earth and
> water and all things are in suspension.
> —Pytheas of Massilia, *On The Ocean*;
> Quoted from Barry Cunliffe,
> *The Extraordinary Voyage of Pytheas the Greek*

PYTHEAS IS BELIEVED TO BE THE FIRST WESTERN EX-plorer to document sea ice, encountering it at some undetermined destination ("Thule") 7 days' sail north of Britain. Pytheas was an astute observer; among other things, he was the first to document the effects of the phase of the moon on tides. While his portrayal of sea ice could be construed as the description of jellyfish, Pytheas was probably trying to find words for thin sheets of young ice. Pytheas made other references to "mare concretum"—the frozen ocean—and he described the midnight sun, so his account is plausible. Sea ice must have been a difficult notion for someone that hailed from the Mediterranean. The original writings of Pytheas no longer exist, so we rely on the interpretations of Pliny, Strabo, and Diodorus. Of course, at this time the Sami and Dorset people were well established on the Arctic coastline, and the Tuniit, predecessors of the Inuit, had been living with the rhythms of the sea ice for more than two millennia, with travel, migration, and hunting tailored around the seasons and moods of the frozen ocean.

This chapter provides a brief introduction to the geography and physics of sea ice. The seasonal flood and ebb of sea ice parallels the advance and retreat of snow and ice over the land surface, with many similarities to the thermodynamics of freshwater ice described in chapter 4. Cold air temperatures drive accretion from below, freezing the seawater. Salinity effects and ocean currents complicate matters in sea ice, however, and the scale of sea ice and its resultant climatic influences are also global in scope. Sea ice has been recognized as a central component in Earth's climate dynamics for several decades now, and most global climate models include a reasonably sophisticated treatment of sea ice.

CHARACTERISTICS OF SEA ICE

Sea ice forms from freezing of seawater. Dissolved salts—predominately NaCl, but many other ions as well—depress the freezing point of seawater by approximately $0.054°C$ ppt^{-1}. For mean ocean water with a salinity of 34.5 ppt, sea ice forms at $-1.86°C$. Polar waters are often fresher than average seawater, due in part to limited evaporation. For salinities of 25 ppt and 30 ppt, water freezes at $-1.35°C$ and $-1.62°C$, respectively.

Figure 5.1. A variety of Arctic sea-ice types and seasons. (a) Ridged landfast ice. (Photograph courtesy of Andy Mahoney, National Snow and Ice Data Center, University of Colorado.) (b) One-week-old ice—*nilas*—forming in Reykjavík Harbor, December 2000. (Photograph by author.) (c) Late-summer ice floes, Tanquary Fjord, northwestern Ellesmere Island, Canada. (Photograph courtesy of J. Dumas.) (d) Pancake ice in the Bering Sea. (Photograph by R. Behn, NOAA Corps.)

First-year sea ice forms in the autumn in the polar regions. New ice has many colorful terms to distinguish it, including nilas, grease, slush, and pancake ice. Figure 5.1 illustrates a spectrum of sea-ice types in the Arctic, including fresh pancake ice in the Bering Sea, late-summer ice floes in the Canadian high Arctic, and a fully developed winter ice cover in the Beaufort Sea (landfast ice on Alaska's North Slope).

Sea ice is made up of a mixture of brine, ice crystals, air, and solid salts. Once nucleated, sea ice provides a platform for the deposition and accumulation of meteoric snow. Brine pockets trapped in first-year sea ice give it salinity values that are commonly in the range 5–15 ppt. During the summer melt season, sea ice reaches the melting point (0°C) and becomes permeable, with water and brine transport along intergranular veins. Brine rejection through this process freshens sea ice. For this reason, multiyear ice, which has survived the summer melt season, has only traces of salinity.

Sea ice thickens through the fall and winter, growing from below through basal accretion, or *aggradation*. Thermodynamic growth of first-year ice is self-limiting to a thickness of about 2 m, or less than this when sea ice is mantled in a thick snow cover. At thicknesses beyond this, the ocean is effectively insulated from cold atmospheric temperatures, and ocean heat flux into the base of the sea ice is balanced by upward heat conduction through the ice and snow. Thicker ice develops through mechanical ridging under convergence, compression, and overriding (rafting) of ice floes. Pressure ridges can reach thicknesses of 10–20 m.

Through the spring and summer in both the Arctic and Antarctic, a large fraction of first-year ice melts away or is advected to lower latitudes. That which survives the summer becomes multiyear ice, which goes through further growth stages (thermodynamically and through ridging), commonly reaching ice thicknesses of a few meters.

Figure 5.2 plots measured and modeled estimates of the winter thickness distribution of Arctic and Antarctic sea ice. The thick multiyear sea ice evident in figure 5.2a is a result of ice convergence against the Canadian Arctic Archipelago and northern Greenland, which gives thicknesses of about 5 m. These data represent a snapshot from February to March 2005 and are derived from satellite altimetry in the Arctic basin. Ice thicknesses vary from year to year, but this general geographic pattern is persistent. In Antarctica, a more radially symmetric ice thickness pattern is evident, associated with ice divergence away from the coast. Thicker ice along the coast indicates multiyear ice in regions of ice convergence, particularly evident in the western Weddell Sea where ice piles against the Antarctic Peninsula.

During the summer melt season, atmospheric heat flux is the primary driver of sea-ice melt, although there is also basal melting from ocean heat fluxes. Melt begins gradually due to the high albedo of the seasonal snow cover. Once the snow ablates, melt ponds form on the ice surface (e.g., figure 1.2), the albedo is dramatically reduced, and melt proceeds swiftly. In the Arctic, more

Figure 5.2. Thickness distribution of winter sea ice in (a) the Arctic basin and (b) the Southern Ocean. (a) Arctic ice thickness from March 2006, derived from ICESat altimetry (Kwok et al., 2009). (Image courtesy of NASA's Goddard Scientific Visualization Studio.) (b) Mean September sea-ice thickness from a regional (circumpolar) configuration of a finite element sea-ice–ocean model. (Image courtesy of Ralph Timmermann, Alfred Wegener Institute.)

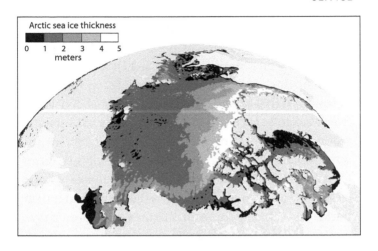

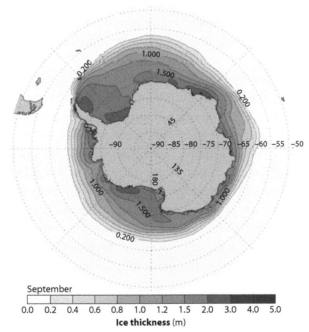

than half of the ice pack typically melts away in the summer, and more than 85% of the Southern Hemisphere sea ice melts each summer (see table 1.1).

Although climate warming is being felt in the Arctic (chapter 9), this seasonal cycle is remarkably consistent from year to year. Figure 5.3 plots Arctic and Antarctic sea-ice extent for the period 1990–2000, based on microwave remote-sensing measurements of monthly mean ice extent. Sea-ice extent refers to the total area with at least 15% sea-ice cover; ice extent is therefore greater than the ice area. This figure testifies to the overall dominance of seasonal insolation cycles in governing sea-ice cover; other aspects of the climate system, such as variability in wind, pressure, and ocean conditions, influence year-to-year ice anomalies but these are difficult to discern in figure 5.3.

The geometry of the Arctic basin also plays a hand in the consistent maximum winter ice extent in the north; winters are cold enough that most of the Arctic basin freezes over each year, with ice extent limited and defined by the continents. Antarctic sea ice is not continentally constrained, but its annual maximum is also very consistent, governed by the "ablation wall" imposed by the relatively warm waters of the Antarctic Circumpolar Current. These plots indicate total ice cover in each hemisphere; regional ice cover is much more variable.

Figure 5.4 plots the geographic extent of sea ice for the times of minimum and maximum ice cover in each hemisphere. The white areas indicate the minimum and maximum ice extents in 2009, and the black lines

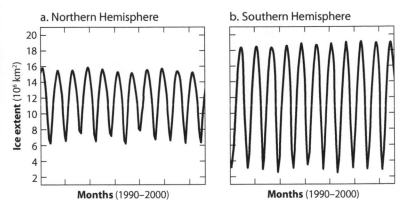

Figure 5.3. Monthly sea-ice extent in the (a) Northern Hemisphere and (b) Southern Hemisphere, January 1990 to December 2000. (Data from the U.S. National Snow and Ice Data Center.)

indicate the median monthly ice extent for the period 1979–2000.

The polar oceans are not covered by a continuous, uniform sheet of ice. Rather, sea ice is made up of a mixture of open water, pack ice, and *landfast* ice, along with the occasional iceberg that has come off of a terrestrial ice mass and been entrained in the sea ice. Sea ice *concentration* refers to the areal fraction of ice cover in a region. Landfast ice, often called fast ice, is frozen to the shore, or it can be anchored to the seafloor in shallow, near-shore environments. It swells with the tides but is otherwise immobile. Ice cover in landfast ice is generally continuous. Open-water ice floes, in contrast, are discontinuous and highly mobile. Pack ice drifts at speeds of order 0.1 m s^{-1}, driven by ocean currents and wind stress

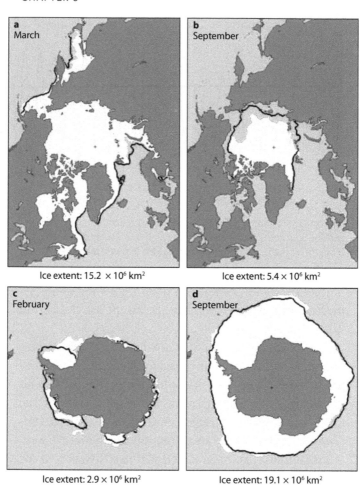

Figure 5.4. Maximum and minimum sea-ice extent in the (a, b) Northern Hemisphere and (c, d) Southern Hemisphere. All plots show the monthly minimum and maximum ice extents in 2009, along with the median value for the period 1979–2000 (black lines). (Data from the U.S. National Snow and Ice Data Center.)

and subject to Coriolis deflection. Pack ice, also called drift ice, consists of a mixture of ice floes and open water, with the latter taking the form of *polynyas* (open areas) or *leads* (cracks between ice floes). Open-water areas are sources of heat and moisture flux to the atmosphere. Many polynyas are persistent features from year to year, as open water is maintained by sea-ice divergence (due to prevailing winds or currents) or upwelling of warm ocean waters.

In the Arctic, pack ice that is entrained in the Beaufort gyre can circulate through the Arctic basin for several years before being exported to the North Atlantic through Fram Strait and the channels of the Canadian Arctic Archipelago. The East Greenland Current is a kind of "sea-ice alley" where huge volumes of sea ice are advected southward each year, including thick, multiyear ice. The redoubtable Fridtjof Nansen recognized this and many other nuances of sea ice during his voyages in the Arctic in the late 1800s. Nansen described this ice upon approach to southeast Greenland in summer 1888, shortly before disembarking to complete the first crossing of the Greenland ice sheet:

It must not be supposed that this drifting ice of the Arctic seas forms a single continuous field. It consists of aggregations of larger and smaller floes, which may reach thicknesses of thirty or forty feet or even more. How these floes are formed and where they come from is not yet known with certainty, but it must be somewhere in the open sea far away in the north, or over against the Siberian coast, where no one has hitherto forced his way. Borne on the polar current, the ice is carried southwards along the east coast of Greenland.

Thick, multiyear ice is prevalent where winds pile up ice in parts of the basin, such as the northern coasts of the Canadian Arctic Archipelago (figure 5.2a). The thick ice floes that Nansen observed drifting southward would have been multiyear ice from the archipelago or the Beaufort Sea gyre that was caught up in the transpolar drift and exported through Fram Strait. The geography is much simpler in Antarctica. Offshore winds from the continent push ice northward, giving a divergent ice pack with little multiyear ice. This offshore export is very effective in melting first-year ice, giving the strong summer minimum in the Southern Ocean that is seen in figure 5.3b.

These processes also give systematic differences in ice concentration in each polar region; the average annual ice concentration in the Northern Hemisphere is 83%, versus 72% in the Southern Hemisphere. In the Northern Hemisphere, the average (1979–2010) ice area varied from 4.8×10^6 km^2 in September to 13.6×10^6 km^2 in March. Ice extents for the same period ranged from 6.6×10^6 to 15.5×10^6 km^2. In the Southern Hemisphere, ice area and extent for the period 1979–2010 ranged from 1.9×10^6 to 14.5×10^6 km^2 and 3.0×10^6 to 18.8×10^6 km^2, respectively, with a minimum in February and maximum in September. It is difficult to estimate hemispheric-scale ice area and extent prior to satellite observations.

Whereas seasonal insolation cycles and land–sea geography shape the sea-ice extent in each hemisphere, interannual and decadal variability in sea-ice thickness and concentration are influenced by air temperature, ocean heat fluxes, ocean circulation, and atmospheric pressure patterns, which drive surface winds. As discussed earlier, prevailing winds can concentrate ice along coastlines, supporting thick, multiyear ice, or they can drive ice divergence and export. Because these meteorological controls are immediate, sea ice adjusts rapidly to climate variability and change. An anomalously warm summer can lead to loss of sea ice, with numerous positive feedbacks that include decreased local and regional-scale albedo, solar radiative heating of open water, increased sensible and longwave heat transfer from open water to the atmospheric boundary layer, and effects of open water on cloud cover. In the Arctic, these positive feedbacks contribute to multiyear "memory" and decadal-scale variability in ice volume. There is less multiyear ice in the Southern Hemisphere, so sea-ice volume in the Southern Ocean has less memory; thermodynamic processes here produce interannual variability but limited decadal variability.

Exceptionally detailed views of ice area, extent, and motion are available for the modern, satellite era (e.g., figures 5.2–5.4), whereas other aspects of sea ice are difficult to quantify. In particular, basin-scale measurements of ice thickness are elusive. Many local observations are available, and upward-directed sonar from submarine surveys provides good transect data, but sea ice is constantly shifting. Repeat sonar surveys along a particular transect provide information about ice-thickness changes over time, but they do not tell the complete story. Interannual variability in drift or convergence of the pack ice can give large differences in thickness in a region even though there may be little or no change in total basin ice volume.

SEA-ICE THERMODYNAMICS

The discussions in chapters 2–4 are salient to many aspects of sea-ice thermodynamics. Brine content influences the thermal properties of sea ice, relative to freshwater ice, but temperature evolution and ice thickness are still broadly governed by vertical diffusion. Heat advection from percolation of brine and meltwater can also be significant.

Similar to lake ice, heavy snow cover on a thin platform of sea ice can submerge the ice, causing flooding that creates "snow ice." Freezing of seawater in these conditions is fundamentally different from basal ice accretion, and it can lead to high levels of salt entrainment. This process is particularly strong in Antarctica, due to pack ice divergence that creates a relatively thin ice cover that is vulnerable to submergence. The crystal structure and radiative properties of snow ice differ from those of accretion ice. Snow ice also has a high thermal conductivity relative to snow, promoting sea-ice growth.

Ocean heat flux, Q_w, is also more variable and potent than the geothermal heat flux into the base of terrestrial ice masses or the basal heat fluxes normally experienced by freshwater ice. Typical values are 1–2 W m^{-2} but can be higher where warm water masses (e.g., 0°C or even a few degrees Celsius) penetrate to high latitudes. This is known to transpire in the Northern Hemisphere through intermittent forays of North Atlantic water into the Arctic basin. Eddies from the Antarctic Circumpolar Current also deliver warm water to coastal Antarctica. Warm water can upwell buoyantly or through Ekman divergence, delivering large heat fluxes to sea ice. A water mass of this type is found in the Southern Ocean and is called *circumpolar deep water*. This warm, salty water has its origins in the North Atlantic overturning circulation.

The thermodynamic formulations described in chapter 3 apply to sea ice or a combined sea ice and snow layer, with some modifications for the effects of salinity content. Combining the sensible and latent heat content of the sea ice, the effective heat capacity, or "thermal inertia," of sea ice with salinity S and temperature T is written

$$c_{si}(T,S) = c_i + \frac{\mu L_f S}{T^2}, \tag{5.1}$$

where c_i and L_f are the specific heat capacity and latent heat of fusion of freshwater ice, S is in parts per thousand, and T is in degrees Celsius. The coefficient μ comes from the linear approximation to the effects of salinity on melting-point depression: $T_m = -\mu S$.

Couched in these terms, one can simulate the energy required to raise the temperature or melt a given volume of sea ice based on its temperature and salinity. For ice density ρ_{si}, integration of (5.1) over temperature gives an equation for the energy per unit volume required to warm a parcel of sea ice from temperature T_1 to T_2:

$$E = \rho_{si}c_i(T_2 - T_1) - \rho_{si}L_f\mu S\left(\frac{1}{T_2} - \frac{1}{T_1}\right). \tag{5.2}$$

Equation (5.2) neglects the temperature dependence of the specific heat capacity of pure ice, although this can be included in numerical models. If $T_2 = T_m$, the melting temperature, this implies a complete phase change from ice to liquid brine. Equation (5.2) can then be written in terms of the total energy required to melt a volume of sea ice with an initial temperature T and salinity S:

$$E_m = \rho_{si}c_i(T_m - T) + \rho_{si}L_f\left(1 + \frac{\mu S}{T}\right). \tag{5.3}$$

This is equivalent to the enthalpy per unit volume of the ice. These equations are stable because $T < 0°C$ for saline ice. For freshwater ice, the energy of melt is $E_m = \rho_i [c_i (T_m - T) + L_f]$.

Ice growth or basal melting, including the effects of brine content, can then be modeled as a function of enthalpy. This is an adaptation of the equation for the growth of freshwater ice given in Eq. (4.6). For ice thickness H,

$$-\rho_{si}E(S,T)\frac{\partial H}{\partial t} = Q_{in}\big|_{b-} - Q_{out}\big|_{b+} = Q_w + k_i\frac{\partial T}{\partial z}\bigg|_{b+}, \tag{5.4}$$

where Q_{in} and Q_{out} refer to the upward-directed vertical heat flux into and away from the ocean–ice interface.

Q_{in} is equal to the ocean heat flux, and Q_{out} is the heat conducted upward into the sea ice. This equation is combined with the surface energy balance at the upper boundary (the snow/ice–atmosphere interface), as described in chapter 3, and the conservation of enthalpy within the ice volume. Substituting enthalpy for internal energy (temperature) in Eq. (3.5), the thermodynamic evolution of sea ice follows

$$\frac{\partial E}{\partial t} = \frac{\partial}{\partial z}\left(-k\frac{\partial T}{\partial z}\right) + \vartheta \tag{5.5}$$

where ϑ accounts for solar radiative heating and latent energy release/consumption from internal refreezing/melting. A term can also be added to account for heat advection from brine percolation, given a model of that process.

To define fully sea-ice thermodynamics, salinity evolution $S(z,t)$ also needs to be accounted for. Along with (5.4) and (5.5), this gives a system of coupled equations for the temporal evolution of sea ice of thickness H, with vertical temperature and salinity structure $T(z)$ and $S(z)$. Similar to the heat advection associated with brine migration, this is difficult to measure or model. At present, most sea-ice models prescribe salinity or hold it fixed.

SEA-ICE DYNAMICS

Sea-ice drift is governed by the Navier–Stokes equations, which describe the force balance in similar fashion to the conservation of momentum equations for circulation of the ocean or atmosphere. Motion is driven by wind stress and ocean currents, subject to Coriolis deflection and mechanical dissipation of momentum through ice interactions. For horizontal ice velocity u,

$$\rho_{si}\left(\frac{\partial u}{\partial t} + u \cdot \nabla u\right) = -\rho_{si} f \times u + \tau_a + \tau_w - \rho_w g \nabla h_w + \nabla \cdot \sigma, \quad (5.6)$$

where ρ_{si} and ρ_w are the density of sea ice and ocean water, f is the Coriolis parameter, τ_a and τ_w are the frictional force from wind and ocean currents, expressed as shear stress per unit length, h_w is the sea-surface height, and σ is the internal stress field in the sea ice. The second-to-last term captures the pressure gradient associated with sea-surface slope, and the final term describes deformation in the ice pack, which extracts energy from the mean flow. In other words, internal deformation, through processes such as rafting, ridging, and shearing of ice floes, take up some of the energy provided by wind and ocean drag, limiting the acceleration and motion of the ice pack. These processes also shape the thickness distribution of the ice pack, with major feedbacks on ice thermodynamics.

In free drift, sea-ice flow is predominantly wind-driven, with Coriolis deflection giving motion that is commonly 30° to 60° to the right (left) of the wind in the Northern (Southern) Hemisphere. Surface water currents in the ocean are also strongly wind-driven, so these frictional forces commonly collaborate in propelling sea-ice motion.

Calculation of internal deformation in sea ice, $\nabla \cdot \sigma$, requires a constitutive relation or rheology that describes how ice deforms under stress. Ice is an unusual solid in that it exhibits elastic, viscous, and plastic behaviors, depending on the timescale and the nature of the stress that is imposed. The slow, viscous (tertiary creep) deformation that is of great interest for glaciers and ice sheets is not of much relevance to sea ice, but elastic flexure and plastic failure are common modes of deformation in shifting pack ice. Sea ice is commonly modeled as a viscoplastic material, which means that it deforms linearly with stress until loading stress reaches the yield strength, beyond which the material fails (i.e., sea ice will fracture and break up). Sea ice is modeled as a continuum, so explicit models of fracturing or shattering are not usually entertained; treating sea ice as a plastic material mimics this decidedly discontinuous process. Plastic deformation can describe sea-ice rifting (opening of leads) and ridging, two of the most important mechanical processes.

Sea ice has also been described as a cavitating fluid, which is weak under tension but has a high yield stress under compression. This can mimic sea-ice behavior, while being simpler computationally then a full plastic rheology. It is also possible to model sea ice as an elasto-viscoplastic material, where elastic deformation is of interest. This is the case in some engineering applications, such as calculation of sea-ice stresses on marine structures (e.g., bridge abutments, drilling platforms). Conical (sloping) abutments are designed to permit sea-ice "ride up" under elastic flexural bending, minimizing stress on the structure. Similar elastic processes are relevant to sea-ice deformation when pack ice floes converge. Sustained convergence inevitably leads to either rafting (one ice floe overriding the other, common in thin young ice) or ridging (failure under compression, leading to crumpling and a thick, jumbled suture zone). Rules are needed to describe ice ridging and the changes in ice thickness distribution and ice-covered area associated with this process.

Internal forces and mechanical dissipation in the ice pack also play an important role in ice thickness distribution. Rafting, ridging, and shearing are the main processes that create thick ice beyond thermodynamic growth limits. This is difficult to model in a fixed-grid, Cartesian framework, so two approaches are commonly adopted. One is to simulate the mean ice thickness in a grid cell, along with the area of ice cover (or areal fraction within the grid cell, also called the compactness). Together these give the ice volume. Converging ice floes as well as thermodynamic growth will increase the mean ice thickness and compactness, subject to conservation of energy and mass (volume). Melting thins the ice pack.

As a refinement to this approach, many models adopt a statistical distribution of the ice thickness distribution in a grid cell (e.g., normal distribution, bimodal distribution, etc.). The probability distribution function for this is denoted $g(H)$, for ice thickness H, as illustrated in figure 5.5. The distribution can then be sampled at discrete intervals in order to simulate sea-ice evolution for different ice-thickness categories. Thermodynamic growth and melting of the ice proceed at different rates as a function of ice thickness, and mechanical effects

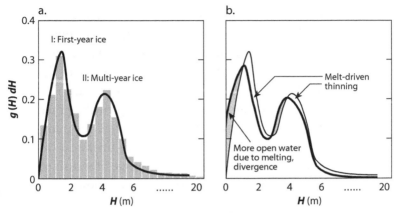

Figure 5.5. Illustration of sea-ice thickness distribution, $g(H)$. [Adapted from Bitz and Marshall (2011).] (a) The line depicts a hypothetical ice distribution in an ice-filled grid cell with a combination of first-year and multiyear ice (modal thicknesses of ca. 1 m and 4 m). The histogram illustrates a potential discretization of $g(H)$ into ice-thickness categories for sea ice modeling. (b) A hypothetical shift in $g(H)$ as ice is melted and open water is created in the region. The thin and thick lines indicate the original and new ice-thickness distributions, respectively, with the shading indicating areas of net "gain" of probability.

create transfers of mass between different ice-thickness categories. There is therefore a governing equation for the probability distribution function that needs to be coupled with (5.6),

$$\frac{\partial g}{\partial t} = -\nabla \cdot (gu) - \frac{\partial}{\partial H}\left(g\frac{\partial H}{\partial t}\right) + \Pi - \Upsilon, \qquad (5.7)$$

where Π describes the evolution of g due to ice-mechanical thickness changes, and Υ represents ice-pack ablation from lateral melting. The first term on the right-hand side includes changes in ice thickness distribution due to both divergence in the ice pack ($g\nabla \cdot u$, for mean velocity u) and advection of the distribution ($u \cdot \nabla g$). The second term on the right-hand side couples Eq. (5.7) with the thermodynamic equation for ice growth or melt, Eq. (5.4). Rates of change, $\partial H/\partial t$, are resolved separately for the mean ice thickness in a specific ice-thickness category, as determined by the method selected for discrete sampling of the probability distribution function.

Further Considerations

Many of the processes associated with sea ice are complicated and difficult to observe, such as brine migration, internal stresses, deformation of the ice pack, the extent of basal melting, and basin-scale ice thickness distributions. However, sea ice is such a recognized, fundamental part of the climate system that most climate models include some treatment of sea-ice dynamics and thermodynamics.

Spatial and temporal patterns of sea-ice variability have many affects on the ocean and atmosphere. Sea ice is unique in its ability to shift seasonally large areas of open ocean to something that is essentially land, thermodynamically and physically. Atmospheric and oceanic conditions at high latitudes cannot be simulated well without a good representation of sea ice. This is also true of biogeochemical and water vapor fluxes between the ocean and atmosphere, with interesting and poorly understood feedbacks on the high-latitude hydrological cycle, cloud conditions, and gas and aerosol exchanges. A full exploration of these topics is beyond the scope of this overview, but I return to some of these processes in chapters 8 and 9.

SUMMARY

Sea ice is a year-round feature of the polar oceans, with seasonal ice cycles shaping the weather, climate, biology, and ecology of the polar regions. In the north, sea ice also shapes the social and cultural rhythms of indigenous people, through its influence on transportation and traditional fishing and hunting practices. Sea ice also extends to midlatitudes in the winter season, affecting shipping in protected near-shore waters such as the North Sea and Baltic Sea, offshore of Japan in the Sea of Okhotsk, and the Cabot Strait waters of eastern Canada.

The thermodynamics of sea ice are well represented by the lake ice processes discussed in chapter 4, with the addition of salinity effects. However, ocean currents and winds make sea ice a considerably more complex aspect of the cryosphere. Advection of ice floes, opening and closing of leads, and mechanical interactions between floes create an inhomogeneous ice cover with large variations in ice thickness and type. Ice rafting, ridging, shearing, and extensional failure require consideration of ice rheology: the way in which sea ice deforms, fractures, and fails under stress. This mechanical dissipation in turn feeds back on the ice thickness distribution, ice thermodynamics, and the momentum balance governing ice drift. All of this introduces multiyear memory in polar sea ice cover, where anomalies in the ice pack and associated ice–climate feedbacks can persist for many years. Chapters 8 and 9 examine aspects of sea-ice-climate interactions in further detail.

···

[One] who keeps company with glaciers
comes to feel tolerably insignificant by and by.
—Mark Twain, *A Tramp Abroad*

GLACIERS ARE PERENNIAL ICE MASSES THAT ARE LARGE
enough to experience gravitational deformation: the
flow of ice under its own weight. Glaciers and ice sheets
nucleate where snow accumulation exceeds snow ablation over a period of many years or decades. With time,
the accumulated snow is buried, compressed, and transformed to glacier ice. Ice behaves as a nonlinear, viscoplastic fluid; once ice thickness is sufficient, internal
gravitational stresses cause the ice to deform.

Mountain glaciers are found throughout the world's
alpine regions, including the high mountains of tropical
East Africa, South America, and Guinea. There are many
types of mountain glacier, classified primarily from the
topographic setting, but prominent among these are
cirque and *valley glaciers*. These can be independent ice
masses, self-sufficient in an alpine niche, or valley glaciers can also be *outlet glaciers* that drain *icefields* or
névés: low-sloping snow accumulation areas that mantle

mountain peaks or high-elevation plateaus. Icefields conform to the topography, with high peaks looming above the ice surface on the icefield perimeter and within the ice mass itself. Peaks or ridges that poke out of the ice are known as *nunataks*. The structure of the mountains and the valleys that separate them dictate ice flow directions and the shape and extent of an icefield. This distinguishes icefields from *ice caps*, which overwhelm the underlying topography, with ice flow determined by the geometry of the ice cap itself. Figure 6.1 provides examples.

An *ice sheet* is a continental-scale ice cap. There is no particular glaciological distinction other than size, but the Greenland and Antarctic ice sheets are more than two orders of magnitude larger than the world's largest ice caps, and they play a unique role in the world's climate, so they warrant special consideration. These ice sheets contain ancient "Ice Age" (Pleistocene) ice, hundreds of

Figure 6.1. Examples of icefields, outlet (valley) glaciers, ice caps, and ice sheets. (a) The Columbia Glacier, an outlet glacier of the Columbia Icefield. (Photograph courtesy of R.W. Sandford.) (b) The Greenland ice sheet, along with icefields and ice caps on Baffin Island, Devon Island, and Ellesmere Island in northeastern Canada. (Image from Google Earth.) (c) Elephant Foot Glacier, a piedmont glacier in northeast Greenland. (Photograph courtesy of the Alfred Wegener Institute.) (d) Antarctica, with the gray line indicating the grounding line; all ice outside of this is floating in the ocean. (Image by Bob Bindschadler/NASA, courtesy of the U.S. Antarctic Program.) (e) Iceberg calving in a proglacial lake from a valley glacier in Kenai Fjords National Park, Alaska. (Satellite image courtesy of GeoEye.)

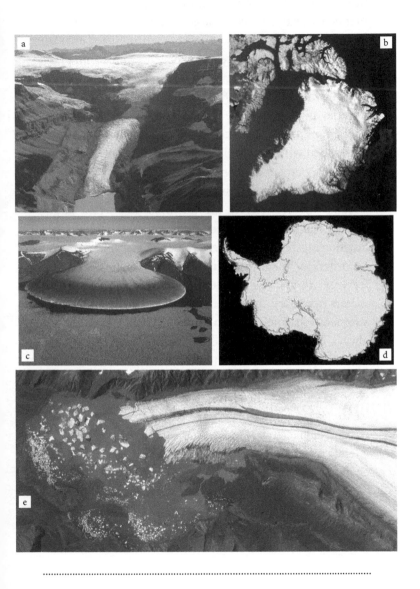

thousands of years old, and they reside in polar environments with extreme cold; large regions of each ice sheet are too cold to experience surface melting. They are also large enough to influence atmospheric circulation, planetary albedo, global sea level, and many aspects of regional climate (chapter 8). These are the only ice sheets in today's world, but the Pleistocene ice sheets in North America, Eurasia, Iceland, and Patagonia, which waxed and waned with the glacial–interglacial cycles, had similar features and effects on the climate.

There are various flow features within ice sheets. Ice streams and ice shelves are the most important large-scale features for ice sheet dynamics. Ice streams are fast-flowing regions of the ice, where ice velocities can be orders of magnitude higher than those in adjacent ice (e.g., thousands of meters per year versus tens of meters per year). Ice streams are normally laterally constrained—channelized in a sense—although they are not always constrained by bedrock topography. Ice stream width in parts of Greenland and Antarctica is enigmatic and varying, related to bed conditions. Warm, wet-based sections of the ice sheet support fast flow, in contrast with ice that is frozen to the bed. We return to this later in discussion of glacier dynamics. In this type of ice stream, the lateral margins are known as *shear zones*: narrow features characterized by heavily fractured ice.

Ice shelves are marginal areas of the ice sheet that are free-floating in the ocean, composed of glacier ice that has flowed out onto the sea (rather than sea ice, which is frozen in situ). Ice shelves rim much of Antarctica, where they are hundreds of meters thick. The large ice shelves are fed by numerous outlet glaciers or ice streams. There are also ice shelves in the Arctic, on Ellesmere Island and Greenland, although these are much smaller and in many cases are considered to be floating ice tongues: marine-based extensions of individual outlet glaciers.

GLACIER AND ICE SHEET GEOGRAPHY

The polar ice sheets are distinguished by their size, but they are also unusual presences on the landscape in many other ways, given to extremes. They create topography that rivals the world's major mountain chains. The summit of the Greenland ice sheet has an elevation of 3207 m, whereas much of the Antarctic ice sheet plateau crests above 4000 m. Both ice sheets are underlain by major mountain ranges, with occasional nunataks, and the range of Transantarctic Mountains that separates East Antarctica and West Antarctica is 3300 km in length. The deepest ice in Greenland is 3370 m thick, and the deepest known ice in Antarctica is 4780 m thick.

Permanent ice covers about 84% of the island of Greenland and 99.7% of Antarctica, and the permanent population at each site scales accordingly; Greenland is home to just more than 56,000 people, most of them clustered in picturesque, ice-free villages around the southern coast of the island, whereas about 1000 people reside year-round at the scientific research bases in Antarctica. There is a rich cultural history in Greenland, dating back several millennia, whereas Antarctica was only imagined prior to the first documented sighting in 1820. Although Antarctica is a scientific treasure, it is not a hospitable environment; it is the coldest and windiest place on the planet. A temperature of –89.2°C has been recorded at Vostok Station (78°27' S, 106°52' E), with an annual average temperature of about –55°C. Katabatic winds that charge down the slopes of Antarctica reach speeds of 250 km hr^{-1}, and the average wind speed at Mawson Station is 37 km hr^{-1}. The vast interior region of the East Antarctic plateau is a polar desert, receiving annual snow accumulations of less than 50 mm w.e.

Glaciologists generally divide Antarctica into three sectors: West Antarctica, East Antarctica, and the Antarctic Peninsula. The division is based on the geography, topography, and dynamics of the ice sheet in each region. West Antarctica is largely marine-based and is thinner and lower in elevation than its eastern counterpart, with an average thickness of 1050 m (1310 m if one excludes the ice shelves). The Bentley subglacial trench, which underlies a portion of the West Antarctic ice sheet, reaches a depth of 2496 m below sea level. East Antarctica sits atop a broad continental craton, resting mostly above sea level even in its current isostatically depressed state. The average thickness of the East Antarctic ice sheet is 2146 m (2226 m excluding the ice shelves), about twice that of West Antarctica.

The Antarctic Peninsula is a maritime setting that extends to a lower latitude than the rest of the continent. The climate and terrain have more in common with Patagonia than with the Antarctic plateau, leading to a more alpine style of regional glaciation. Icefields on the peninsula are drained by outlet glaciers that terminate near the coast or flow into ice shelves that fringe the peninsula. Because of their relatively low latitude and altitude, glaciers and ice shelves of the Antarctic Peninsula are the only ice masses on the continent to experience surface melting in most years. Melting in the Greenland ice sheet is more widespread but is still confined in most years to a steep, narrow zone at the ice sheet periphery, making up about 20% of the entire ice sheet.

These ice sheets constitute most of the world's glacier ice, by area and by volume (table 6.1). They are mysterious and poorly observed parts of the Earth climate system in many ways, although they are arguably better monitored than the ca. 200,000 mountain glaciers, icefields, and ice caps that decorate the world's mountain and polar regions. The vast majority of these are unnamed and unstudied.

Mountain glaciers are numerous but relatively small. As an example, in the mid-1970s there were an estimated 5154 glaciers covering 2909 km^2 in the European Alps, an average glacier size of 0.56 km^2. This is of course a snapshot in time; in 1999, the glacier area in the Alps was 2416 km^2, comprising 5422 individual glaciers, giving an average glacier size of 0.45 km^2. Methods for counting glaciers differ between studies. Data is derived from satellite imagery, aerial photos, and maps from different periods, and one must make choices about the minimum size for an ice body to be considered a glacier, and how to divide ice fields into multiple, discrete outlet glaciers. Glacier

Table 6.1

Global Glacier Area and Volume

Region	Area (10⁶ km³)	Volume	
		10⁶ km³	msl
Antarctica			
East Antarctic ice sheet	9.86	21.7	51.6
West Antarctic ice sheet[a]	1.81	3.0	4.6
Ice shelves	1.62[b]	0.7	—
Peripheral icefields and ice caps[c]	0.47	0.17	0.43
Total	13.76	25.6	56.6
Greenland			
Greenland ice sheet	1.68	2.93	7.1
Peripheral icefields and ice caps	0.06	0.03	0.07
Total	1.74	2.96	7.2
Rest of the world			
Dyurgerov and Meier (2005)	0.54	0.13	0.31
Ohmura (2004)	0.52	0.05	0.13
Raper and Braithwaite (2005)	0.52	0.09	0.23
Radić and Hock (2010)	0.52	0.16	0.41
Global total	16.03	28.6	64.1

Note: msl is meters of global eustatic sea level equivalent, correcting for ice that is floating or grounded below sea level (see the section "Glacier and Ice Sheet Volume").

[a] West Antarctic values exclude the Antarctic Peninsula.

[b] Includes ice rises.

[c] Includes the Antarctic Peninsula, estimated at 300,400 km² and 95,000 km³ (0.24 msl).

retreat since the 1970s has reduced glacier extent, but the basic characteristics of the glacier distribution hold true, with more than 80% of glaciers in the Alps less than 1 km² in size. A 2005 inventory of glaciers in western Canada (south of 60°N) gives an estimate of 17,600 glaciers covering 26,730 km^2, with an average size of 1.5 km^2. Glacierized area in western Canada declined by 11% relative to a similar snapshot for 1985, but the number of glaciers is stable or increasing due to fragmentation as icefields thin and retreat.

Larger icefields and ice caps are found at high latitudes in both hemispheres. Most of these have their origins at high elevations, from which they spread out over the terrain. Glaciers descend to sea level in locations like Alaska, Iceland, Patagonia, and the Arctic and Antarctic islands. The largest midlatitude icefield in the world is the Southern Patagonian ice cap, covering 16,800 km^2 over a latitude range 48.3°S to 51.5°S. This icefield is notoriously harsh: windswept with annual precipitation totals exceeding 10 m w.e. Ice cores indicate annual accumulations as high as 17 m w.e. in some years. Few glaciological studies have been done here, for obvious reasons—the region makes the weather in coastal Alaska and Norway seem hospitable.

The extent of terrestrial ice in the world's main glaciated regions (excluding Antarctica and Greenland) is listed in table 6.2. Arctic Canada harbors the greatest quantity of ice, with the Prince of Wales Icefield on Ellesmere Island (19,325 km^2) representing the world's single largest ice mass outside of Greenland and Antarctica. The network of alpine icefields straddling the Alaska–Yukon border in the Saint Elias, Wrangell, Chugach, and northern Coast Mountains spans an estimated 88,000 km^2, but it is heavily dissected so is difficult to compare with the large ice caps in the Canadian

Table 6.2
Major Glaciated Regions Outside of Greenland
and Antarctica

Region	Ice Area (10^3 km^3)
Canadian Arctic	147
High-mountain Asia[a]	114
Alaska	75
Western Canada	53
Svalbard	37
Himalayas	33
Novaya Zemlya	24
Patagonia	21
Severnaya Zemlya	19
Franz Josef Land	14
Iceland	11

[a] Includes Himalayan glaciers.

Arctic. It is probably possible to traverse more than 20,000 km^2 in the Saint Elias Mountains without removing one's skis.

There are about 116 ice caps in the world outside of Greenland and Antarctica. These make up only 0.05% of the world's population of small glaciers, but they represent about half of the ice in the small-glacier reservoir. The total area of the ice masses outside of Greenland and Antarctica is about 530,000 km^2 (table 6.1). Peripheral ice masses in Antarctica and Greenland that are discrete and dynamically independent of the large ice sheets cover an additional 450,000 km^2, including the icefields of the Antarctic Peninsula.

The glacier areas in tables 6.1 and 6.2 represent snapshots from the mid-1970s, largely based on Landsat imagery from the period 1972–1981. Glacier area is declining as a result of climate warming, as noted above for the Alps and for western Canada, but this has not yet induced major changes to the areas of the Greenland and Antarctic ice sheets, which dominate the global ice area. Extrapolating from glacier changes in the midlatitude mountain regions of North America and Europe, the worldwide decrease in ice cover from the mid-1970s to the 2000s is of order 5000 km^2. Compared with a global ice area of 16×10^6 km^2, this has a negligible influence on planetary albedo, particularly relative to changes in sea ice and seasonal snow cover. The marked glacier decline in mountain regions does, however, affect regional-scale water resources, climate, alpine ecology, and global sea level.

GLACIER AND ICE SHEET VOLUME

Glacier area can be well mapped from satellite imagery and aerial photographs, but global ice volume is more difficult to estimate. Mapping of ice thickness requires surface or airborne radar surveys. Electromagnetic wave frequencies of 5–50 MHz are typically used for glacier depth sounding, as these relatively low frequencies limit signal attenuation and allow penetration of radar pulses through several hundred meters of glacier ice. Resolution at these frequencies is of the same order as the wavelength, 6–60 m for the range 5–50 MHz. The ice–bed interface provides a strong electrical contrast, so these systems have good success in mapping subglacial topography and ice thickness. Other geophysical techniques (e.g., seismic and gravity surveys) have also been applied to ice-thickness mapping.

Detailed (1–5 km) survey grids have been flown over the Greenland and Antarctic ice sheets, providing the basis for ice-volume estimates in table 6.1. Directly translated, the ice volumes in the Greenland and Antarctic ice sheets are equivalent to 7.4 and 63.8 msl (meters of global eustatic sea level equivalent), respectively. Eustatic sea level is defined as the mean height of the world's oceans relative to the fixed surface of the geoid, assuming global ocean area is fixed.[1] In reality, portions of the ice sheets are grounded below sea level, particularly in West Antarctica, and the ice shelves are floating; if this ice were to melt, it would not contribute to global sea level rise. The actual global sea level rise associated with complete, instantaneous loss of all of the ice in Greenland and Antarctica would give a sea level rise of about 64 m: 7.2 msl from Greenland and 56.6 msl from Antarctica.

This is approximate, as ocean basin area changes as a function of water volume. In addition, both the seafloor and the continents experience elastic and long-term viscous (isostatic) responses to changes in water and ice loading. The subglacial bedrock in much of Greenland and Antarctica has been depressed by more than 1000 m. Where the bed lies below sea level in Antarctica, removal of the ice sheet would cause initial flooding of the area—essentially an expansion of the Southern Ocean.

Over centuries to millennia, isostatic relaxation would decrease the volume of the subglacial cavities and contribute to additional sea level rise.

Instantaneous removal of the part of the West Antarctic ice sheet (WAIS) that is susceptible to marine ice-sheet instabilities would give a global eustatic sea level rise of 3.2 m. Elastic rebound of the deglaciated WAIS bed would contribute an additional 0.06 msl, and the long-term (e.g. 10-kyr) isostatic adjustment gives a further 0.4 msl. Large portions of Antarctica and Greenland are also grounded below sea level: 41% of the Antarctic ice sheet, including most of West Antarctica, and 22% of the Greenland ice sheet. Deglaciation of either ice mass would expose extensive marine basins. In East Antarctica and Greenland, most of these would isostatically rebound to elevations above mean sea level over a timescale of centuries to millennia. There are important regional exceptions to this, where deep marine channels incise into the interior of each ice sheet.

Outside of Greenland and Antarctica, ice thickness measurements have been made on several dozen individual valley glaciers and ice caps. The Icelandic ice caps, in particular, are exceptionally well mapped, but this is unusual; glacier volume is poorly known in most other regions. Several approaches have been applied to estimation of ice thickness or volume based on other aspects of glacier geometry. Three common methods include (i) estimation of ice thickness as a function of distance from the ice margin, based on ice rheology and including assumptions about the subglacial topography (e.g., a flat or uniformly sloping bed), (ii) volume–area scaling, and (iii) estimates of local ice thickness from surface slope.

Of these methods, volume–area scaling is the most simple and popular method for estimation of glacier volume. Ice-volume data from approximately 100 glaciers around the world give the empirical relationship

$$V = cA^{\gamma}, \tag{6.1}$$

where $V(10^6 \, \mathrm{m}^3)$ is the volume and $A(\mathrm{km}^2)$ the surface area of the glacier. The parameters c and γ require regional calibration, and it is also common to adopt different values of the scaling parameters for different glacier sizes. Empirical data for the worldwide distribution of glaciers give the exponent $\gamma = 1.36$. A power-law scaling relationship between glacier area and volume has also been theoretically derived, based on the rheologically determined surface profiles and aspect ratios of steady-state valley glaciers and ice caps. This gives $\gamma = 1.375$ for valley glaciers, and the classical parabolic geometry of ice caps corresponds with $\gamma = 1.25$.

Volume–area scaling is believed to be generally applicable to a large ensemble of glaciers. Individual glaciers can deviate by more than 50% from the aggregate relationship as a result of complex or deeply eroded bed topography, unusual ice flow regimes (e.g., extensive glacier sliding), or as a result of being far out of equilibrium. The relationship is not intended to be used on individual glaciers.

Estimates in table 6.1 for the volume of ice locked up in the world's mountain glaciers and ice caps are primarily based on volume–area scaling. Differences stem from contrasting assumptions about the scaling-law coefficients, glacier size distributions, the distinction in form between glaciers and ice caps, and uncertainty in estimates of glacier area. Excluding the peripheral ice-fields in Antarctica and Greenland, the estimates span a range from 51×10^3 to 164×10^3 km^3, which translates to a eustatic sea level equivalent of 13–41 cm. There is an additional, poorly constrained area and volume of ice in the peripheral ice caps in Greenland and Antarctica; including the Antarctic Peninsula, this may amount to an additional 40–60 cm of global sea level equivalent. Combining the minimum and maximum estimates for different regions gives an estimated 56–97 cm of total eustatic sea level equivalent in global ice masses that are dynamically independent of the Greenland and Antarctic ice sheets.

As an alternative to volume–area scaling, first-order approximations of ice thickness can be made from the surface slope, based on the empirical observation that glacier ice has a gravitational driving (shear) stress, τ_d, of about 100 kPa at the base. This is a result of ice rheology once again; ice deforms through nonlinear viscous flow in a way that is predictable, with an effective viscosity that allows glacial ice to support about this amount of shear stress. Glaciers thicken until they reach values close to this, and further thickening or steepening leads to increased ice flow, self-regulating to give $\tau_d \approx 100$ kPa. Taking advantage of this relationship, it is possible to estimate local ice thickness, H, through the equation

$$H = \frac{\tau_d}{\rho_i g \nabla s},$$ (6.2)

where ρ_i is the ice density, g is gravity, and ∇s is the surface slope. The relation is local by definition, and it breaks down when surface slopes become small (or zero), as occurs in the accumulation area of large ice-fields and ice sheets. Average glacier slope is sometimes applied in this case.

This relation provides rough estimates of ice thickness, but there are large local and regional exceptions to the "100 kPa rule of thumb." Steep, deep parts of Greenland's Jakobshavn Isbrae ice stream, for instance, have shear stresses of 200–300 kPa. In contrast, the low-sloping Siple Coast ice streams in West Antarctica operate at shear stresses of 20–40 kPa. In this situation, basal flow prevails over internal deformation of the ice, so (6.2) is not valid. Like volume–area scaling, then, inferences from surface slope only provide a rough approximation of ice thickness.

The fact that we only know the volume of the world's mountain glaciers to a factor of two is humbling with respect to our knowledge of the planet, but it also points to an obvious target for advances in Earth system observation.

GLACIER MASS BALANCE

Mass Balance Processes

Glaciers have relatively simple requirements to be viable: snow accumulation must exceed ablation. Snow accumulation occurs primarily as meteoric snowfall (derived from atmospheric precipitation), but snow can also accumulate at a site through wind deposition or avalanching. Ablation refers to the loss of snow and ice through melting, sublimation, wind erosion, and *calving*, a process whereby slabs of ice at the glacier margin mechanically fracture and detach from the main ice mass. This is an effective ablation mechanism for glaciers and ice sheets that are in contact with the ocean, where iceberg calving removes large amounts of ice. Most melting occurs at the glacier surface—the ice–atmosphere interface—but there is also melting internally (englacially), at the glacier bed (subglacially), and on vertical ice cliffs at the glacier margin, particularly where glaciers reach the sea and a large area of ice can be in contact with water. Melting only leads to ablation in the case where meltwater runs off and is removed from the system; some surface meltwater in the upper regions of glaciers and ice sheets percolates into the snowpack or ponds at the surface, where it can refreeze.

Glacier mass balance is a measure of the net accumulation minus ablation over a specified time interval, typically 1 year:

$$B = A_c - A_b, \tag{6.3}$$

where A_c and A_b denote accumulation and ablation. B represents the rate of change in mass or volume of an ice body (a glacier or ice sheet), with units kilograms per year or cubic meters per year. It is also common to see mass balance described at a point on the glacier, or averaged over a glacier surface. Strictly speaking, this is the *surface mass balance*, B_s, which includes snow accumulation and melt but neglects basal and internal accumulation and ablation, as well as mass loss associated with calving. For most mountain glaciers, surface mass balance dominates total glacier mass balance, so $B_s \approx B$.

When defined at a point, surface mass balance is best expressed as a specific mass balance, which is a measure of the gain or loss of mass per unit area of the glacier,

$$b_s = a_{cs} - a_{bs}. \tag{6.4}$$

Here, a_{cs} and a_{bs} refer to the local snow accumulation and ablation (primarily melt) at the glacier surface, with units kilograms per square meter per year or meters water equivalent per year (m w.e. yr^{-1}). This is traditionally measured in glacier mass balance programs through stake and snowpit data. Snow density measurements are required to convert changes in snow-surface height to changes in mass. Averaged over the area of a glacier, this gives the mean specific surface mass balance, $\bar{b}_s = B_s/A$, for glacier area A. This is a useful measure of the average rate of thinning or thickening of the ice mass. In the literature this is often referred to as the net mass balance, b_n, but it is only a measure of surface mass balance, not the total glacier mass balance.

Given several years of positive mass balance, snow is buried, compressed, and transformed to firn and then glacial ice. Glacier dynamics transports ice from areas of net accumulation to low elevations, bringing perennial ice to places where local climate is too warm or dry to support glacier cover otherwise. This lower part of a

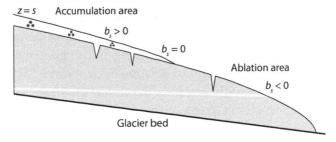

Figure 6.2. Schematic of a valley glacier, indicating the accumulation and ablation areas and the end-of-summer snowline, generally called the equilibrium line altitude (ELA). b_s refers to the annual surface mass balance (accumulation – ablation).

glacier, known as the ablation area (figure 6.2), is characterized by bare ice at the end of the melt season. Figure 6.1c provides a good example. Above this lies the accumulation area. The line separating the two is called the equilibrium line altitude (ELA), although this terminology is misleading. It is really the end-of-summer snowline, the elevation at which snow accumulation balances ablation in a given year, and it does not speak to the state of equilibrium of a particular glacier. For a given glacier, however, there is a unique value of the end-of-summer snowline, ELA_0, which corresponds with a net annual mass balance of zero. A glacier will evolve into a state of equilibrium if the condition $\text{ELA} = \text{ELA}_0$ is sustained for many decades.

Glacier mass balance processes differ in tropical, mid-latitude, and polar regions. It is not possible to give a comprehensive account here, but it is a reasonable generalization to say that the mass balance of most mountain glaciers is governed by the surface mass balance (snowfall vs. surface melt), whereas other forms of ablation can be dominant in polar regions. Surface melting accounts for about 50% of the annual ablation in Greenland. Calving of icebergs to the ocean—"dynamical discharge"—is also important in Greenland, where it accounts for 40% to 50% of Greenland's annual mass loss. The remaining ice loss is associated with melting at the ice–ocean interface. Almost all of the ablation in Antarctica occurs through calving of icebergs at marine margins and basal melting in ice shelves and floating ice tongues. West Antarctica has major ice shelves in the Ross Sea and Weddell Sea (figure 6.1d), as well as active floating glacier outlets in the Amundsen Sea, making the WAIS exceptionally sensitive to marine influences.

Climate Sensitivity

Because most of the world's glaciers have been out of equilibrium since the peak of the Little Ice Age in the late 1800s, it is now common to find relict glacial ice in places where the climate once supported glacier growth, but it has now become too warm. In these cases, glacier ice is no longer nourished through net snow accumulation or ice transport. This anomalous ice is slowly melting away, but it can remain on the landscape for decades or centuries while this occurs, particularly if it is insulated by a thick debris cover.

There is no simple "threshold temperature" for a glacier to be viable. A mean annual temperature below 0°C is not a necessary or sufficient condition for glacier ice to exist. Valley and outlet glaciers that are in contact with the ocean, known as *tidewater glaciers*, are vivid examples of this. Because ice flow delivers ice to low elevations, glaciers can extend to sea level environments where mean annual temperatures are several degrees above 0°C. Although most glaciers do not reach the ocean, this feature is intrinsic to mountain glaciers; glacier ice in the ablation area can extend to relatively mild environments, particularly where there is ample snowfall.

It is common to distinguish between *maritime* glaciers, which live in mild, wet climates, and *continental* glaciers, characterized by cool, dry conditions and low mass turnover. Interannual variability in glacier mass balance is sensitive to both annual precipitation and melt-season temperature, but the latter is the stronger control in most settings, particularly for continental glaciers. For instance, in the Canadian Rockies it requires an increase in snow accumulation of about 70% to offset a warming of 1°C. The climatic sensitivity of glaciers differs in maritime environments such as New Zealand, Svalbard, and Norway, where snow accumulation increases of 20% to 30% are needed to offset a warming of 1°C.

Fluctuations in mountain glaciers are not perfectly synchronous in different regions, due to differing regional patterns of precipitation and temperature variability. As an example, increases in precipitation in northwestern Europe during the positive phase of the North Atlantic Oscillation (NAO) deliver extra snowfall to coastal Norway, promoting a more positive mass balance. This is usually accompanied by reduced moisture delivery to southern Europe. As another example from western North America, El Niño–Southern Oscillation (ENSO) causes opposite mass balance anomalies in mid-latitudes (i.e., the Cascade Range and Coast Ranges of Washington State and British Columbia) and the sub-polar latitudes of coastal Alaska.

There are also local-scale (i.e. individual glacier) differences within a region due to the specific topographic setting of a glacier, the way in which this influences local climate conditions (e.g. topographic shading, wind redistribution), and the dynamics of a particular ice mass. Bedrock and glacier hypsometry influence the extent of an ice mass that is affected by a shift in the ELA. This, the general mass balance regime, and the overall glacier size affect the response time of a glacier to a mass balance perturbation.

Several concepts have been introduced to describe glacier response to climate change. Glacier *reaction time* refers to the time delay between a climatic forcing, such as years of positive or negative mass balance, and the ensuing response of the glacier terminus: ice-margin advance or retreat. In mountain glaciers, terminus positions react to mass balance perturbations that are maintained for several years, with typical time lags of about 10 years. This can be quantified by looking at the lagged correlation between annual time series of glacier length and surface mass balance.

The concept is useful but it is poorly defined, because a given glacier may be in a state of advance or retreat at the time when a particular climate perturbation occurs, with a different reaction time expected in each case. A glacier's reaction to changes in temperature or precipitation may also vary, as these impact the glacier differently (i.e., in the ablation vs. accumulation area). The timescale for the dynamical propagation of these changes to the ice front depend on the nature of the climate perturbation, the glacier length, and the flow speed. Further, fluctuations of the terminus (glacier length) can also occur as a result of variability in ice dynamics. Glaciers may speed up in response to increases in meltwater or other environmental influences. Some glaciers undergo cycles of advance and retreat that are driven by internal dynamical cycles rather than climatic forcing (e.g., surging behavior or tidewater instabilities). This can lead to anomalous rates of either advance or retreat that are not synchronized with atmospheric forcing, so fluctuations in surge-type and tidewater glaciers need to be interpreted with care. These glaciers are still subject to long-term climatic control, but dynamical excursions are superimposed on climatic adjustments.

Glacier *response time* offers an alternative measure. It is better defined than reaction time, but it is a theoretical construct. It is defined as the e-folding time[2] for a glacier that is in equilibrium to come to a new state of equilibrium following a step perturbation to climate or mass balance. This is a more intrinsic measure of a given glacier's dynamical response time to climate forcing and should be repeatable or predicable for a given glacier geometry, flow characteristics, and mass balance regime. It is simple to calculate in a model but is difficult to measure or calculate from field observations, as real glaciers are rarely in a state of equilibrium. It nevertheless provides some insight into the timescale of glacier response to climate change, which turns out to be of the order several decades for most mountain glaciers. This means that most glaciers no longer have a strong memory of the Little Ice Age, but they are still responding to climate conditions from the second half of the 20th century. In contrast, large ice caps and ice sheets have response times of centuries to millennia. In Greenland and Antarctica, part of the present-day ice sheet evolution consists of "secular trends" related to the deglaciation from the last glacial maximum, which began about 20,000 years ago.

GLACIER DYNAMICS

Glaciers flow through three different mechanisms: internal "creep" deformation, decoupled sliding at the ice–bed interface, and deformation of subglacial sediments (figure 6.3). The former is a function of ice rheology and the stress regime in the ice, and the latter two mechanisms are governed by conditions at the base of the glacier. These processes are described in more detail below.

Governing Equations for Glacier Dynamics

Similar to models of atmosphere, ocean, and sea-ice dynamics, the flow of glaciers and ice sheets is mathematically described from the equations for the conservation of mass, momentum, and energy. For a point on the glacier with ice thickness H, the vertically integrated form of the conservation of mass is

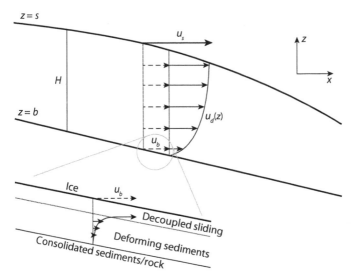

Figure 6.3. Schematic of glacier flow mechanisms. The surface velocity $u_s = u_b + u_d(s)$. Basal velocity is the sum of deformation of underlying sediments and decoupled sliding at the ice–bed interface. Where ice is moving at the bed, these two processes can operate together (lower image), or only one of them may be active. High water pressure facilitates both processes.

$$\frac{\partial H}{\partial t} = -\nabla \cdot (\bar{u}H) + b_s, \tag{6.5}$$

where $\bar{u}$ is the average horizontal velocity in the vertically integrated ice column, and b_s is the specific mass balance rate at a point. The first term on the right-hand side describes the horizontal divergence of ice flux, and the second term describes the net local source or sink of mass associated with accumulation and ablation. The vertically averaged velocity includes ice flow due to both internal deformation and basal flow: $\bar{u} = \bar{u}_d + u_b$. Glacial ice moves slowly, so a year is typically adopted as the most convenient unit of time; hence, ice velocities are reported in units meters per year, and b_s in (6.5) is expressed as meters per year of ice-equivalent gain or loss of mass.

The main challenge in modeling glaciers and ice sheets is evaluation of the velocity field. Acceleration and inertial terms are negligible in glacier flow, so the Navier–Stokes equations that describe conservation of momentum reduce to a case of Stokes flow, where gravitational stress is balanced by internal deformation in the ice:

$$\nabla \cdot \sigma = -\rho_i g. \tag{6.6}$$

Here, σ is the ice stress tensor, ρ_i is ice density, and g is the gravitational acceleration. Several texts present detailed derivation of this full system of equations and their solution.

As for sea ice, a constitutive relation is needed to express internal stresses in terms of strain rates in the ice. $\nabla \cdot \sigma$ can then be rewritten as a function of the three-dimensional (3D) ice velocity field, providing a framework to solve for $\bar{u}$ and integrate Eq. (6.5) to model the evolution of glacier geometry in response to variations in ice dynamics or climate. Because the timescales and stress regimes in glaciers and ice sheets are different from those in sea ice, the mechanical properties and modes of deformation of interest are distinct from those discussed in chapter 5. The next section describes the constitutive relation that is most commonly used in glacier modeling.

....

Ice rheology is strongly temperature dependent, so an additional equation is needed to solve for the 3D temperature distribution. The local energy balance gives the governing equation for temperature evolution in the ice sheet,

$$\rho_i c_i \frac{\partial T}{\partial t} = \rho_i c_i v \cdot \nabla T - \frac{\partial}{\partial z}\left(k_i \frac{\partial T}{\partial z}\right) + \vartheta. \tag{6.7}$$

Here, v is the 3D velocity vector, c_i and k_i are the heat capacity and thermal conductivity of ice, and ϑ represents strain heat production due to deformational work, $\vartheta = \sigma_{ij} \dot{\varepsilon}_{ij}$, where $\dot{\varepsilon}$ is the strain rate tensor. This is the same form as the general equation governing the internal temperature evolution of snow or sea ice (Eq. 3.5), with the addition of an advection term that accounts for heat transfer due to movement of the glacier ice. The advection term includes both horizontal and vertical flow, which are of comparable magnitude for heat advection in glaciers. Only the vertical component of diffusive heat transport is retained in (6.7), because vertical gradients in temperature are much larger than horizontal temperature gradients in glaciers and ice sheets. The solution to (6.7) is subject to prescription of air temperature on the upper boundary (the glacier surface) and geothermal or ocean heat flux at the base of the ice.

Glaciers in mild environments are *isothermal* or *temperate*; summer temperatures and the latent heat release from refrozen meltwater are high enough to give a mean annual surface temperature of 0°C. A winter cold layer may penetrate the upper ~10 m of the glacier, but the underlying ice is at the pressure melting point, $T_{\text{pmp}} =$

$-\beta H$, where $\beta = 8.7 \times 10^{-4}$ K m^{-1} in glacier ice (see chapter 2). Impurities can further lower the melting-point temperature. Unless a glacier is very thin, the bed is insulated from the air temperatures, and geothermal heat flux provides a source of heat energy that warms the base of the ice.

Cold or *polythermal* glaciers are found in sub-Arctic, Arctic, or high-elevation environments where the mean annual air temperature is far below freezing. These ice masses may be frozen to the bed or they may be warm-based (at the melting point). Many polar icefields, including the Greenland and Antarctic ice sheets, have mixed conditions: regions where ice is frozen to the bed and regions where ice is warm-based, with cold ice above (e.g., figure 6.4). Much of Antarctica is warm-based because the ice is extremely thick and has been there a long time, which has allowed the slow trickle of geothermal heat hundreds of millennia to warm things up. Once warmed to the melting point, additional heat input generates basal meltwater, which has ponded in parts of Antarctica to give large subglacial lake basins.

Péclet numbers in polythermal glaciers and ice sheets are of order 1; advection and vertical diffusion are similar in magnitude. This gives an interesting thermal structure where the upper half of a glacier or ice sheet is typically cold, strongly influenced by diffusion and vertical advection of surface conditions (mean annual air temperature). At the ice divide in ice caps and ice sheets, advection toward the bed—"downwelling"—gives colder temperatures throughout the ice column; central Greenland is

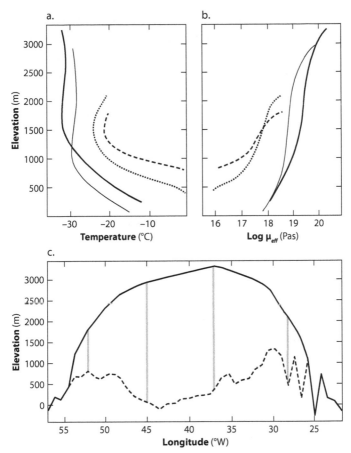

Figure 6.4. Variation of (a) internal temperature and (b) effective viscosity with depth at points along (c) an E–W transect through the Greenland ice sheet at 72.6° N. The heavy solid lines are at the ice divide [Greenland Ice Core Project (GRIP) ice core location, 38.6° W], thin solid lines are along the western flank (45° W), heavy dashed lines are near the western margin (52.2° W), and thin dotted lines are near the eastern margin (28.2° W). Locations along the ice-sheet transect are indicated in (c).

cold-based for this reason (figure 6.4). Horizontal advection of this "cold plume" is evident on the flanks and margins of the ice sheet, where a cold tongue of ice can be sandwiched between warmer ice both above (from warmer atmospheric conditions) and below. Near the ice sheet margins, high rates of shear deformation produce strain heating that supplements geothermal fluxes. This is effective in warming the ice near the bed. Rates of ice flow and the thermal diffusivity of ice make heat transfer in glaciers and ice sheets a slow process. It takes decades for a change in temperature in the atmosphere to penetrate to the base of a glacier, and tens of thousands of years in an ice sheet.

Given a 3D temperature distribution through the ice sheet, the effective rheology of the ice can be evaluated and the velocity field can be numerically determined. Knowledge of the temperature field is also essential to assessing whether the base of an ice mass is at the pressure melting point or not; if so, liquid water can be present at the bed, and the glacier or ice sheet is subject to basal flow.

Internal Deformation

To solve the momentum balance, ice sheet stresses need to be expressed as velocity fields, via a constitutive relation for ice. The rheology of polycrystalline glacier ice is well studied in laboratory and field environments, which reveal that ice deforms as a nonlinear viscous fluid. The original form of the flow law, proposed by John Glen and John Nye in the 1950s, is broadly supported by field studies of tunnel and borehole deformation, as well as observations and modeling of large-scale ice motion. This constitutive relation is known as Glen's flow law:

$$\dot{\varepsilon}_{ij} = f(T, \sigma'_{ij}) \sigma'_{ij}, \tag{6.8}$$

where $\dot{\varepsilon}_{ij}$ is the strain rate, T is ice temperature, and σ'_{ij} is the deviatoric stress tensor in the ice. The flow law is an empirical relation rather than a physical law, à la Newton, but it is rooted in the theoretical assumption that strain rates in ice should be a function of the stress tensor and its invariants. The deviatoric stress is used in (6.8) due to the observation that ice deformation is independent of confining pressure (normal stresses).

For a linear (Newtonian) fluid,

$$f(T) = 1/\mu(T), \tag{6.9}$$

for the viscosity μ. In glacier ice, the effective viscosity, μ_{eff}, is represented as a function of the second invariant of the deviatoric stress tensor, $\Sigma'_2 = (\sigma'_{ij} \sigma'_{ji})^{1/2}/2$,

$$f(T, \sigma'_{ij}) = 1/\mu_{\text{eff}} = B(T) \Sigma_2'^{(n-1)}. \tag{6.10}$$

$B(T)$ is an "ice softness" term that follows an Arrhenius temperature dependence,

$$B(T) = B_0 \exp\left(-\frac{Q}{RT}\right). \tag{6.11}$$

B_0 is called the Glen flow-law parameter, R is a constant, and Q is the creep activation energy. Ice deformation is typically modeled as an $n = 3$ process, giving

$$\dot{\varepsilon}_{ij} = B(T) \Sigma'^{2}_{2} \sigma'_{ij}. \tag{6.12}$$

This formulation is an isotropic flow law that allows the first-order effects of ice temperature and deviatoric stress regime to be incorporated in estimates of ice deformation. Where shear stress and shear deformation are dominant, as is often the case, this is well approximated by

$$\dot{\varepsilon}_{xz} = B(T) \sigma'^{3}_{xz}. \tag{6.13}$$

Glen's flow law is for pure, isotropic ice. There are numerous other complicating factors for ice deformation, such as anisotropic ice fabric, the potential impact of grain size, and the presence of impurities and intergranular liquid water content. These effects are not explicitly resolved in ice sheet models, so the flow rate parameter, B_0, is typically tuned to approximate the bulk effects of crystal fabric, grain size, and impurity content.

Even without this level of detail or complexity in modeling ice rheology, there is tremendous variability in the effective viscosity of ice associated with the range of ice temperatures and stress regimes found in Earth's glaciers and ice sheets. Figure 6.4b plots the effective viscosity variation with depth at sample location in the Greenland ice sheet, as calculated from (6.10).

The strain rates in (6.12) or (6.13) can be expressed as velocity gradients and then vertically integrated or inverted and substituted into the momentum balance (6.6) to give a set of equations for the horizontal ice velocity. Various numerical solutions to these equations have been adopted in glacier and ice sheet modeling.

Basal Flow

In addition to the internal deformation described above, ice can flow at the base where the bed is at the pressure melting point, through some combination of subglacial sediment deformation and decoupled sliding over the bed. Large-scale basal flow generally requires pressurized subglacial meltwater, which can lubricate the bed, float the ice, or weaken subglacial sediments. Subglacial hydrology plays a pivotal role in fast-flowing glaciers and ice streams. High subglacial water pressures can decouple the ice from the bed by reducing or eliminating basal friction. On local scales this may not entice a significant ice-dynamical response, as resistive stresses can be taken up at adjacent well-coupled regions of the bed, by side drag from valley walls or adjacent ice, or by longitudinal stress bridging (upstream and downstream resistance to flow). However, numerous observational studies report occasions where inputs of surface meltwater to the bed overwhelm these resistive stresses and produce localized speedups in both valley glaciers and polar icefields.

For large-scale ice stream flow or surging of outlet glaciers, subglacial water must occupy a significant portion of the glacier bed, at pressures that are sufficient to drown geologic and topographic pinning points. In this situation, widespread ice–bed decoupling can permit high rates of basal flow (hundreds to thousands of meters per year) and a regime in which ice fluxes are dominated by basal flow. In glaciers that exhibit high rates of basal flow, there is uncertainty about the relative importance of sliding along the ice–bed interface versus deformation of the underlying glacial sediments. High subglacial water pressures are conducive to both processes. Fast flow in West Antarctica's Siple Coast ice streams appears to be associated with plastic failure of a thin layer of saturated marine sediment, and similar processes are expected to be important wherever subglacial sediments and topographic features offer a relatively smooth, low-friction substrate. Such flow also appears to have been important in the Quaternary ice sheets in North America and Europe.

Glacier models make some allowance for basal flow, usually through a local sliding "law" relating basal flow, u_b, to gravitational shear stress at the bed, raised to some power m: $u_b \propto \tau_d^m$. Models are oversimplified though, often "switching on" basal motion wherever the glacier bed is at the pressure melting point. This is a necessary but not sufficient condition for large-scale basal flow; many ice masses in the world are isothermal or warm-based but do not experience significant basal motion. In these cases, the glacier may be well drained or the bedrock geology and topography offer too much frictional resistance.

The effects of subglacial water pressure on basal flow have been introduced in some modeling studies, typically through the effective pressure p_e. This is the difference between glaciostatic (ice) pressure and subglacial water pressure: $p_e = p_i - p_w$. An effective pressure of zero indicates that ice is floating, so low or negative effective pressures promote ice–bed decoupling and enhanced basal flow. Although it is safe to assume that $u_b \propto p_e^{-k}$, for some unknown power k, there is likely no generalized local relationship between u_b and p_e; actual basal flow is affected by regional-scale ice dynamics, not just local conditions.

A prescription of the form $u_b = A\tau_d^m/p_e^k$ is unstable as this blows up as $p_e \to 0$. Local flotation is commonly observed in nature, however; for instance, boreholes drilled to the glacier bed can be artesian, creating fountains of water at the glacier surface once the borehole connects with the subglacial water system. Hence $p_e \leq 0$ is a physically acceptable possibility. The mathematical instability is simply a failure of the local form of the basal flow law. An alternative is to introduce a parameterization in terms of the flotation fraction p_w/p_i, with $u_b = 0$ when $p_w = 0$ and basal flow increasing with p_w/p_i. The local expression $u_b = A\tau_d^m f(p_w/p_i)$ can represent this. However, basal flow observations are notoriously difficult to make so there is no clear recommendation as to the functional form of $f(p_w/p_i)$. Hydrological enabling of basal flow is a nonlinear, threshold process. Furthermore, basal flow is not an intrinsically local process, but arises in association with regional stress balance (e.g., longitudinal stresses that propagate upstream from the calving face of a tidewater glacier).

General Discussion of Glacier Dynamics

The equations of motion describe a nonlinear viscous fluid with an effective viscosity that is intermediate between that of liquid water and that which characterizes Earth's mantle; ice deforms slowly, but measurably. Where there is basal friction, vertical shear deformation is the dominant type of strain in glaciers and ice sheets. One implication of the nonlinear flow law is that the bulk of the shear deformation is concentrated near the bed. Another consequence is that the surface velocity (or column-integrated ice velocity) associated with shear flow is a highly nonlinear function of ice thickness and gravitational driving stress (surface slope): $\bar{u}_d \propto H^4 \nabla s^3$. The thermal structure of glaciers reinforces this and can provide a positive feedback on shear deformation, as strain heating warms and softens the ice near the bed. In some cases (e.g., Jakobshavn Isbrae), this creates the development of a thick, strongly deforming temperate layer in the basal ice. Stiff, brittle surface ice that is −40°C can overlie a 1000-m-thick layer of relatively ductile ice at the pressure melting point.

In floating ice, where basal shear stress vanishes, ice deforms through longitudinal spreading, with essentially a free-slip condition at the base because the underlying water offers little frictional resistance. This flow is resisted by longitudinal stresses in the ice and, in most situations, horizontal shear stresses ("side drag") from the walls of a fjord or embayment. These additional stresses are also active in the inland part of ice sheets, where deformation is dominated by vertical shear flow, but are often secondary in this setting. Where ice–bed coupling is weak, as occurs in many ice streams, longitudinal and horizontal shear stresses assume an important role. They are also important in regions of complex ice dynamics, such as near the grounding line (the transition from grounded to floating ice), at ice divides, and in valley glaciers, where velocity and thickness gradients are steep and side drag from valley walls is significant.

Surface velocities of tens of meters per year are typical of the interior regions of the polar ice sheets and in most alpine glaciers. Thick outlet glaciers in steep valleys experience flow rates of 100 m yr^{-1} or more via internal (vertical shear) deformation. Basal motion is usually at play wherever glaciers and ice sheets have higher flow rates than this, such as Antarctic ice streams and in glacier surges. The main exceptions to this are for floating glaciers and in outlet glaciers that occupy deep fjords. Spreading flow rates in floating glaciers and ice shelves are often hundreds of meters per year. Where grounded glaciers discharge through fjords, as occurs in major outlet glaciers of the Greenland ice sheet, deep channels have been carved by glacial erosion, giving glaciers that are exceptionally thick and steep. This promotes high rates of internal deformation and surface velocities that reach thousands of meters per year.

Most of these fast-flow situations are problematic for glaciological models. The physical controls of basal flow are not well understood or parameterized, and deep, narrow fjord environments are poorly resolved in continental-scale models, which typically operate at spatial resolutions of 10–50 km. This is a pressing challenge for ice sheet models that aim to describe interannual- and decadal-scale variability in outlet glacier dynamics and iceberg discharge (and the associated sea level rise) in Greenland and Antarctica. It also means that some important mechanisms of temporal variability, such as glacier surge cycles, rarely arise naturally in glacier and ice sheet models.

....

SUMMARY

One of the wondrous paradoxes of the global cryosphere is that feather-like snowflakes can come together over time to build landscape features as dramatic and massive as the Antarctic ice sheet. Glaciers and ice sheets are a merger of meteorology and geology: a creation of the atmosphere that takes on a solid permanence on timescales of centuries to millions of years. One of the unsettling aspects of recent climate change is the evidence that mountain glaciers and polar ice sheets may not be as permanent and ponderous as previously thought. Through atmospheric and oceanic forcing on glacier mass balance, surface meltwater effects on glacier dynamics, and positive feedbacks involved in both mass balance and ice-dynamical processes, glaciers and ice sheets can be surprisingly responsive to climate variability.

I have dedicated extra pages to the discussion of glacier and ice sheet dynamics, as this is a rich topic that still begs greater understanding with respect to Earth's climate system. The research community is in early stages in its representation of glaciers and ice sheets in climate models, but there is increasing awareness that this element of the global cryosphere has an important role to play in future changes in climate and sea level. I elaborate on this important topic in chapters 8 and 9, with an overview of the role that ice sheets have played in climate dynamics throughout Earth history.

In the bleak midwinter
Frosty wind made moan,
Earth stood hard as iron,
Water like a stone
—Christina Rossetti,
"In the Bleak Midwinter"

PERMAFROST IS HIDDEN BENEATH THE SURFACE AND less in the public consciousness than sea ice, glaciers, and ice sheets, but it directly affects human and biological systems that live with it. Permafrost can be hundreds of meters thick and tens to hundreds of millennia in age, with a geologic intransience but recent changes that remind us that it is not so permanent a feature of the landscape as it seems. Timescales of thermal diffusion in permafrost mean that surface temperature changes take a long time to propagate to depth, similar to glaciers. Seasonally frozen ground and the upper layers of permafrost do respond to seasonal and short-term climate and land-use fluctuations, however, so permafrost processes have major effects on the infrastructure, hydrology, ecology, and carbon cycle of northern high latitudes. This chapter gives a brief overview of permafrost.

PERMAFROST GEOGRAPHY

Permafrost is perennially frozen ground. More specifically, the International Permafrost Association defines permafrost as ground that maintains a temperature at or below 0°C for at least 2 years. Frozen water need not be present for soils or rocks to meet this definition. For practical (engineering) purposes and in the context of cryospheric science, however, it is the ice content of permafrost that is of interest. This requires slight modifications to the formal definition, as we are primarily concerned with whether or not water in the pore space of soils and rock will be in the liquid or solid phase. This depends on the local pressure melting point and the salinity of the pore fluid. As discussed in chapter 2, overburden pressure depresses the freezing point of pure water by 0.074°C MPa^{-1}, giving a freezing point of about −2°C under 1 km of sediments.

In marine environments, pressure melting point and salinity effects are both relevant. The latter dominate on continental shelves, where typical ocean salinities give a freezing point of −1.8°C. Marine sediments with seawater in the pore space need to dip below this temperature to create permafrost. In deep waters (most of the world's oceans), pressure effects require temperatures below −3°C to support permafrost. Water temperatures this cold are not found in the modern ocean, so marine permafrost is only viable in cold, continental shelf environments. These are found in high-latitude continental shelves that were exposed to the atmosphere due to lower sea levels during the Pleistocene glaciations. Once frozen, salt rejection creates low-salinity ice in the sediments, with a melting point closer to 0°C; this helps to preserve frozen ground on continental shelves of the Arctic basin, as they typically see ocean temperatures below this.

About 58% of the Northern Hemisphere land mass, an area of 55×10^6 km^2, experiences seasonally frozen ground. This thaws in the spring and summer in much of this area, but perennially frozen ground—permafrost—covers 23×10^6 km^2, or about 24% of the land in the Northern Hemisphere. Almost half of this is present as continuous permafrost in the high latitudes of Russia, Canada, and Alaska (figure 7.1), defined as areas with permafrost covering more than 90% of the landscape. South of this one encounters zones of discontinuous (50% to 90%) and sporadic (<50%) permafrost. Pockets of alpine permafrost are also present in most mountain ranges, and additional, generally unmapped areas of perennially frozen ground occur under glaciers and ice sheets at high latitudes.

Permafrost is rare in the Southern Hemisphere, because the only place sufficiently cold—Antarctica—is mostly ice covered. About 0.3% of Antarctica is free of glacial cover, much of this in the Dry Valleys. This entire area contains permafrost. An unknown but potentially large area of frozen ground underlies cold-based sectors of the Antarctic ice sheet. Permafrost is also found on the Antarctic islands and at high elevations in the Andes.

An additional, poorly mapped area of permafrost resides in shallow marine shelf environments. Much of this

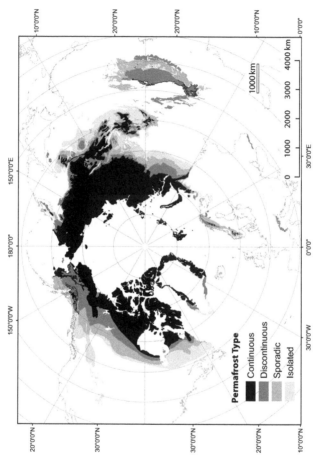

Figure 7.1. Permafrost distribution in the northern hemisphere, based on the data compilation by Brown et al. (1998, updated 2001), available from the U.S. National Snow and Ice Data Center.

formed during the Pleistocene glaciations, when global sea level was drawn down more than 100 m due to the buildup of ice sheets on land. This exposed large areas of continental shelf around the world, and permafrost had thousands of years to develop in these exposed-shelf environments fringing the Arctic Ocean, such as the Bering Sea.

Much of the Earth's frozen ground formed during the Quaternary glacial periods and has persisted to this day. Relict permafrost features can be found at lower latitudes and in mountain regions, where a relatively thin layer of frozen ground developed at these locations during the glaciation but has since thawed. This helps to demarcate the proglacial zone of the last great ice sheets.

Subsea permafrost is more than 100 m thick in places. In general, permafrost thickness ranges from decimeters to more than 1000 m, with the deepest permafrost found in parts of Siberia, Alaska, and northwestern Canada that eluded glacier ice for much of the last glaciation. The greatest known depth of permafrost is 1500 m, found near the Lena River in Siberia. Where ice sheets persisted for extended periods they insulated the ground from cold air temperatures, moderating the mean annual temperature and limiting permafrost growth.

PERMAFROST THERMODYNAMICS

Like lake, river, and sea ice, permafrost forms due to subzero surface temperatures that induce freezing from above. Surface tension effects retain some liquid water content to low temperatures (−10°C or less), particularly in fine sediments such as clays and silts. However, the 0°C isotherm in the ground generally represents the freezing front: the depth at which the phase change begins and ice can be found in the soil or rock matrix. Seasonal ground frost is widespread in midlatitudes and high latitudes, penetrating to a depth of a few decimeters. Where temperatures are warm enough, this frost recedes each spring or summer. In places where the mean annual surface temperature is below 0°C and seasonal frost survives the summer thaw, the freezing front in the ground propagates to depth through thermal diffusion.

The resulting temperature gradient in the ground is typically negative: Temperature increases with depth, and there is an upward-directed conductive heat flux out of the ground. This cooling effect at the base of the permafrost is opposed by geothermal heat flux, as well as other potential heat sources at depth (e.g., heat advection through water transport). Where upward heat conduction exceeds the geothermal heat flux at the base of the permafrost, there will be permafrost growth, or *aggradation*, at the base. When the permafrost begins to melt from above, surface temperatures warm, temperature gradients in the ground lessen (or the ground can become isothermal, at the pressure melting point), and the conductive heat flux in the permafrost is reduced. Geothermal heat flux then drives melting, or *degradation*: thinning from below. Because this is a diffusive process through frozen ground that can be hundreds of meters thick, permafrost growth and decay have very long timescales: decades to tens of thousands of years. This can

Table 7.1
Typical Thermal Properties of Common Near-Surface Geologic
Materials

Soil or Rock Type	Density $(kg\ m^{-3})$	k $(W\ m^{-1}\ °C^{-1})$	c $(J\ kg^{-1}\ °C^{-1})$	κ $(10^6\ m^2\ s^{-1})$
Ice	917	2.1	2090	1.11
Fresh snow	200	0.08	1880	0.22
Settled snow	300	0.21	1880	0.37
Water	1000	0.54	4184	0.13
Air	1.2	0.025	1010	20.63
Quartz	2660	8.8	800	4.14
Clay minerals	2650	2.92	900	1.22
Silt	1600	2.51	1297	1.21
Organic material	1300	0.25	1920	0.10
Clay soil, $q = 0.2$	1800	1.18	1250	0.52
Sandy soil, $q = 0.2$	1800	1.8	1180	0.85
Peat soil, $q = 0.4$	700	0.29	3300	0.13
Icy peat	900	2.09	1670	1.39
Gravel	2000	1.26	750	0.83
Granite	2700	2.51	795	1.17

Note: k, c, and k are the thermal conductivity, specific heat capacity, and thermal diffusivity, respectively, and q is the liquid water content.

be quite well described through one-dimensional (1D) thermal diffusion in many settings.

Other influences on the energy balance such as heat advection from water transport and the warming effects of water bodies (lakes, marine environments) can be locally important in causing lateral variability in permafrost thickness. Thawed ground adjacent to a river or lake is known as a *talik*. The hydraulic permeability of frozen ground is low, so movement of water usually occurs in unfrozen ground beside and below the permafrost or in fractures that can be produced by thermal contraction or desiccation of the ground.

Growth of permafrost occurs through freezing of interstitial water at the freezing front. Low capillary and vapor pressures can lead to migration of water and water vapor to the freezing front, and this promotes the growth of *massive ice*: ice lenses that are millimeters to decimeters thick. These produce frost heave and unusually high ground ice content, so upon melting they lead to large-scale ground subsidence.

The geologic material plays a large role in the response time and depth of permafrost, mostly through controls on the liquid water content. Table 7.1 lists the thermal properties of some common soils and rocks in permafrost terrain. Porous, saturated materials with high water content have a high effective heat capacity, due to the latent energy associated with the advance or retreat of the freezing front. Crystalline rock is at the other end of the spectrum, with high thermal conductivity and a minimal amount of free water in cracks and fractures.

To model thermal diffusion in permafrost, it is common to treat the ground as a saturated matrix of soil, rock, or sediment with ice or water in the pore space, as relevant. For porosity θ and a matrix density ρ_r with ice-filled pores, the bulk density is $\rho_b = (1 - \theta)\,\rho_r + \theta\rho_i$. This framework can also be applied where there is massive ice, which does not occupy the pore space proper, with θ denoting the mass fraction of the ice content in the substrate. Other bulk thermodynamic properties (conductivity, heat capacity) can be calculated accordingly, although it is common and physically better justified to calculate the bulk thermal conductivity through a harmonic weighted average. The equation governing 1D thermal diffusion in the ground is then of the same form as (3.5), but using bulk thermal properties,

$$(\rho c)_b \frac{\partial T}{\partial t} = \frac{\partial}{\partial z}\left(-k_b \frac{\partial T}{\partial z}\right), \tag{7.1}$$

with upper surface temperature prescribed as a boundary condition. A second equation is needed to model the movement of the freezing front. Similar to the growth of sea or lake ice, this is described by the rate of freezing or thawing at the front, $\dot{m}$, following

$$\rho L\theta \dot{m} = Q_{in} - Q_{out} = -k_{bu} \frac{\partial T}{\partial z}\bigg|_{f-} + k_{bf} \frac{\partial T}{\partial z}\bigg|_{f+} \tag{7.2}$$

where Q_{in} and Q_{out} represent the heat flux into and away from the freezing front (W m^{-2}), modeled through heat conduction in the unfrozen ground below and frozen ground above (subscripts u and f, respectively). The vertical position of the freezing front is defined to be $z=f$, with the minus and plus signs denoting the temperature gradients (hence, heat fluxes) below and above this front. Where $\dot{m}$ is positive, there is melting at the front and permafrost thinning. Where $\dot{m}$ is negative, there is permafrost aggradation. Q_{in} can be estimated from the local geothermal heat flux or heat flow in the unfrozen ground underlying the permafrost can also be modeled explicitly, as per (7.1). If appropriate, an advection term accounting for heat transport from water can also be added to (7.1) and (7.2).

Numerical solution of (7.1) and (7.2) is straightforward. A moving grid is recommended to follow the migration of the freezing front, as used in the example of lake ice growth in chapter 4. One common assumption is to assume a thermal steady state in (7.1), which is equivalent to a linear temperature profile in the permafrost; given the upper surface temperature, geothermal heat flux, thermal conductivity of the frozen ground, and the fact that the lower permafrost boundary will be at the pressure melting point, the temperature gradient in the permafrost is then uniquely defined, and permafrost depth can be calculated. This is reasonable for first-order approximations of permafrost thickness, but it cannot account for transient climate effects during permafrost growth and decay.

Another modeling strategy, with some parallels to modeling sea-ice growth and decay, makes use of an effective heat capacity that includes latent heat in (7.1). This is warranted when the freezing front is not sharply defined, as occurs in clay-rich sediments that have significant unfrozen water content at subzero temperatures. For unfrozen water content θ_w, representing the mass fraction, this is modeled from

$$\rho_b \left(c_b + L_f \frac{\partial \theta_w}{\partial T} \right) \frac{\partial T}{\partial t} = \frac{\partial}{\partial z} \left(-k_b \frac{\partial T}{\partial z} \right). \tag{7.3}$$

Here, the second term on the left-hand side represents the latent energy required to melt or freeze free water as temperature changes in the soil. When $\theta_w = 0$ (i.e., at sufficiently cold temperatures, or below 0°C for some geologic materials that support a sharp phase transition), this expression collapses to (7.1). The advantage of (7.3) is that there is no need to track explicitly the freezing front via (7.2); this is implicit in $\theta_w(z)$, and one can define the permafrost thickness purely from the condition $T \leq 0°C$ in the modeled temperature profile. However, (7.3) requires knowledge of the soil water content and its behavior with temperature, which are difficult to measure and simulate.

The methods for modeling subsurface temperature and phase-front evolution are reasonably well established, but it can be surprisingly challenging to predict surface temperature forcing for permafrost. Mean annual surface temperature is not the same as the mean annual near-surface air temperature, which is what is usually measured at meteorological stations. Vegetation cover and snow depth are the greatest influences on this. Both have an insulating effect that gives mean annual surface (ground) temperatures several degrees Celsius above the mean annual air temperature in most situations. There can be exceptions to this. For instance, extremely thin snow covers (e.g., less than 10 cm) do not offer effective insulation. Snow cover that persists into the summer can also cause cooling of the ground relative to mean annual air temperature, as a result of snow-albedo effects and the high thermal emissivity of snow.

The Active Layer

In contrast to the deep permafrost, the temperature and depth of the surface *active layer* are annually varying. This is the top layer of the ground, which is subject to annual freeze–thaw cycles. The active later is of order 1 m deep in most environments, depending on the ground thermal conductivity, the annual temperature cycle, the surface type, and the amount of snow cover. In some locations, it can be several meters deep. Settings with high soil water and organic content have a high heat capacity in the near-surface layer, providing a thermal buffer that limits active layer depth.

Because the active layer is directly forced by atmospheric conditions (in particular, temperature and snowfall), it is exceptionally sensitive to short-term climate change. Atmospheric warming or increased snowfall lead to warmer mean annual ground temperatures and increases in active layer depth. This results in permafrost degradation from above. Warmer surface temperatures also decrease the temperature gradient in the ground, leading to reduced conductive heat fluxes and permafrost decay at the base. When permafrost degrades, then, it commonly warms throughout, and it melts from both above and below. Relict permafrost—"buried ice"—is found at depth in many places, testimony to this process and to sustained periods of colder temperatures in prior times.

Subglacial Permafrost

Cold-based glaciers support subglacial permafrost growth, although glacier cover insulates the ground and results in thinner permafrost than would be otherwise expected at high latitudes or elevations. Permafrost likely preceded ice sheets on the landscape during the last glaciation, and after the ice advanced the ground could warm to the melting point, but no further. This would allow permafrost to persist, but it would likely degrade from below in this situation. Subsurface permafrost is frequently found in the forefield of retreating glaciers in subpolar and polar environments. If subglacial permafrost becomes isothermal, an interesting situation is possible where free water (generated from strain heating, for example) can exist at the base of a glacier in thermal equilibrium with the overlying ice and underlying permafrost. This permafrost interface would be impermeable in this situation, promoting high water pressures at the glacier bed. Such a situation may have contributed to the low-sloping, fast-flowing ice lobes along the southern margin of the Laurentide and Eurasian ice sheets, via high rates of glacier sliding.

Clathrate Hydrates

Frozen ground also hosts an unusual family of substances known as clathrate or gas hydrates. These consist of gases trapped in a crystalline "shell" or "cage" of ice. The ice crystals are connected via customary hydrogen bonds (see chapter 2), but the pressure of the gas molecules inside the shell prevents them from collapsing into a standard ice crystal lattice.

Gas hydrates are only stable under a specialized, sometimes narrow, range of temperature and pressure conditions. Most common are methane (natural gas) hydrates.

These are particularly common in shallow continental shelf environments. Large deposits of methane hydrate have been mapped on the seafloor, in continental-shelf sediments of the Arctic basin, and in terrestrial permafrost, of great interest for their potential as hydrocarbon energy sources. Many other gases are also found in hydrates, including CO_2. The release of CH_4 or CO_2 to the atmosphere through destabilization of gas hydrates as ocean temperatures warm is a potential "wild card" in the climate system, capable of producing abrupt climate change if large quantities of methane hydrate were to melt. It is not clear how much of a threat this is, as seafloor methane is likely to be oxidized as it passes through the ocean water column, and the different reservoirs of methane hydrate are unlikely to destabilize simultaneously.

Permafrost Thermometry

In addition to direct interactions with the climate system, permafrost has been successfully used as a decadal- and century-scale thermometer of climate change. Boreholes in frozen ground record the surface temperature history, based on the diffusion of surface temperature signals to depth. Such records have been examined extensively in recent decades, and they add to the body of evidence with respect to 20th century warming. Borehole temperatures in Alaska indicate a recent warming of up to 4°C. In other locations, borehole temperature records show negligible change or cooling. These records are accurate reflections of ground surface temperature history, but they can be difficult to interpret because of the effects of changing vegetation and snow cover, which can cause a decoupling of the air and ground temperature signals. Thermal diffusion smooths out temperature variability, so it is difficult to invert borehole temperature records for millennial-scale temperature trends.

SUMMARY

Permafrost is a major presence in northern landscapes and is also found in high mountain regions and ice-free parts of Antarctica. Deep permafrost formations probably represent the oldest ice on the planet. Where present, permafrost alters the water table, shapes the regional geomorphology, and creates significant challenges for infrastructure. Some of the highest rates of coastal erosion in the world are found in ice-rich sediments on the Arctic coast, where sea level rise and diminished sea ice are exposing the coast to increased ocean swell. Other aspects of permafrost interaction in the climate system include its potentially significant role in the global carbon cycle through long-term storage of organic carbon, which, if melted, decomposes to release methane and CO_2 to the atmosphere. I discuss this further in chapter 8.

8 CRYOSPHERE–CLIMATE PROCESSES

....

> Nature chose for a tool, not the earthquake or
> lightning to rend and split asunder,
> not the stormy torrent or eroding rain,
> but the tender snow-flowers
> noiselessly falling through
> unnumbered centuries.
> —John Muir, *Studies in the Sierra*

JOHN MUIR'S MUSINGS ON THE EROSIVE POWER OF GLA-
ciation and its legacy in the landscape are a reminder
of the surprisingly potent role that the cryosphere can
play in the Earth system. The Pleistocene glaciations
are a grand example of this: a time when ice defined the
planet and shaped the global climate. These episodes tell
us a great deal about cryosphere–climate processes and
feedbacks, which are still active, if more subtle, in today's
world. This chapter describes the influence of the cryosphere in the climate system.

SNOW AND ICE ALBEDO

Snow and ice are the most reflective natural surfaces on
the planet, exceeded only by clouds in their influence on

....

planetary albedo. This reduces the net solar radiation that is available to warm the Earth, lowering the mean annual temperature of the planet. The cryosphere also introduces regional and seasonal variability in the absorbed solar radiation that is available.

Surface albedo effects of the cryosphere have the greatest impact on regional and local climate through cooling and through influences on atmospheric circulation. The seasonal evolution of snow albedo and the transition from seasonal snow to ice cover on sea ice and glaciers create strong increases in ablation as the melt season progresses. Darkening of the surface causes a three- to four-fold increase in the net shortwave radiation that is available for melt, as illustrated in the surface energy balance example of chapter 3. This feedback drives greater rates of melt in the late summer, contributing to the lag between peak insolation and the annual sea-ice minimum (September in the Arctic and March in Antarctica). This same lag causes runoff from glaciers and ice sheets to peak in late summer.

Seasonal snow cover is the most variable feature of Earth's land surface, with albedo feedbacks serving to amplify the solar-driven seasonality at midlatitudes and high latitudes. In the winter months, insolation is weak or absent at high latitudes, muting the effects of snow cover; the influence of snow and ice on local energy budgets is therefore greatest at midlatitudes.

The impact of snow and ice on the planetary energy budget can be estimated with a number of assumptions. A simple representation of the global energy balance is possible through an adaptation of the linearized, zonally averaged surface energy balance explored in the 1960s by Budyko and Sellers. We consider this here by discretizing Earth into 5° latitude bands, with the zonal mean topography, land fraction, and land/ocean surface properties calculated for each latitude band, θ_j. Each zonal band has a distinct area, albedo, and total heat capacity, A_j, α_j, and C_j ($J\,m^{-2}\,K^{-1}$).

Incoming solar radiation, $Q_s^{\downarrow}$, can be estimated from the potential direct radiation, and the temperature can be solved in each band, following

$$C_j \frac{\partial T_j}{\partial t} = \left[f_j Q_{sj}^{\downarrow}\left(1-\alpha_c\right)\left(1-\alpha_{sj}\right) - \varepsilon \tau_{aj}\sigma T_j^4 - k\left(T_j - \bar{T}_s\right)\right] \quad (8.1)$$

Here, T_j is the zonally averaged surface temperature, and the globally averaged surface temperature is denoted $\bar{T}_s$. In the solar radiation term, f_j is a geometric factor that represents the effective area for incident solar radiation, including zenith angle effects (cf. Eq. 3.9). The factor $(1 - \alpha_c)$ accounts for atmospheric backscatter, primarily associated with clouds, and α_{sj} is the surface albedo. For ice-free terrain, surface albedo is assigned from representative values for the ocean (0.1), forest (0.2), tundra (0.25), and desert (0.3). Snow and ice cover are assigned albedo values of 0.8 and 0.6, respectively. The transition from open water/bare land to sea ice and snow cover is based on a simple temperature rule, with a linearly increasing snow/ice cover as temperature decreases from a mean annual value of 0°C to –10°C. Surface albedo α_{sj} is then calculated as a composite of the ice-free and ice-covered values over this transition. This is simplistic, of course, but this model is purely illustrative.

The second term on the right-hand side describes the net outgoing longwave radiation (Eqs. 3.10 and 3.11), for emissivity ε, including a factor τ_{aj} that represents the atmospheric transmissivity to longwave radiation. This term was linearized in the original energy balance models, but it is simple to use the full nonlinear form of Stefan–Boltzmann's law with an iterative solution. The final term in (8.1) is a crude representation of poleward heat transfer, parameterized as a linear function of the meridional temperature gradient. This roughly describes the role of the Hadley cells, western boundary currents, baroclinic eddies, and so forth, in transporting energy from the tropics to high latitudes.

This one-dimensional energy balance model is no replacement for a global climate model, but it is a rough tool to examine the planetary energy budget and the influence of the cryosphere on Earth's albedo and temperature. Parameters in (8.1) can be tuned to give a mean temperature of 14.0°C for the planet, representative of the current mean state. The planetary albedo, α_p, is equal to 0.32 in this case, including the compound effects of both the atmospheric and surface reflectivities. In a scenario where all of the snow and ice is removed from the planet, $\alpha_p = 0.26$, and the global average temperature increases to 18.2°C (figure 8.1). A simulation of the last glacial maximum (LGM) conditions with this model, with ice sheets extending to midlatitudes, shifts the results to $\alpha_p = 0.35$ and a global average temperature of 8.8°C.

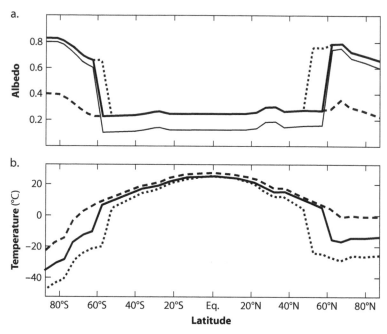

Figure 8.1. Steady-state zonally averaged (a) albedo and (b) temperature in the simple energy balance model of Eq. (8.1), for the present world (heavy solid lines), ice-free world (dashed lines), and LGM world, with continental ice extending to 45° N (dotted lines). The albedo plots in (a) include both atmospheric and surface albedo. The thin solid line plots the surface albedo for the present-day reference model.

These scenarios lack many important climatic feedbacks that would attend such major shifts in climate, such as changes in cloud cover and atmospheric circulation, but the calculations provide a rough estimate of the importance of the cryosphere to planetary albedo.

Similar hypothetical experiments have been carried out with global climate models. In one such study, simulations with LGM boundary conditions give a global cooling of 3°C accompanied by an increase in planetary albedo from 0.31 to 0.33. Application of the same model to an ice-free world gives the expected reduction in surface albedo, but this is offset by an increase in evaporation and cloudiness in this model, giving a warming of 1.3°C but no net change in planetary albedo.

This result demonstrates the sensitivity of the climate system to feedbacks from both the cryosphere and the hydrological cycle. These also have direct connections at high latitudes, where loss of sea ice exposes more open water, increasing local and regional cloud cover. The inferred changes in planetary albedo are massive in terms of the planetary energy balance. At present, average incoming radiation at the top of the atmosphere is $Q_{S0} = 341$ W m^{-2}, of which 102 W m^{-2} is reflected back to space ($\alpha_p = 0.30$), and the remaining energy, $Q_S^{\downarrow} = 239$ W m^{-2}, is available to Earth's surface and atmosphere. This gives $\partial Q_S^{\downarrow}/\partial \alpha_p = -3.4$ W m^{-2} %$^{-1}$. As an example, a loss of snow and ice causing a decrease in planetary albedo from 0.30 to 0.28 would represent an energy surplus of 6.8 W m^{-2}, which compares with an estimated (ca. 2006) radiative forcing of 2.6 W m^{-2} associated with the anthropogenic greenhouse-gas buildup in the atmosphere. In other words, albedo feedbacks can be of a similar or greater magnitude to the greenhouse-gas forcing that is driving current climate change.

185

THE CRYOSPHERE AS A LATENT ENERGY BUFFER

Melting snow and ice requires a great deal of energy. These phase changes consume sensible heat and radiative energy that would otherwise go into warming up a region. The opposite effect attends the fall freeze-up: latent energy is released to the atmosphere, lakes, rivers, oceans, and soil as ice forms from water and vapor. This delays the seasonal decrease in temperatures. Overall then, latent energy exchanges within the seasonal cryosphere act as a thermal buffer, similar to the way in which ocean heat capacity moderates the climate of marine environments.

The amount of energy involved in these phase changes can be estimated from the extent of the seasonal cryosphere, although the area of the seasonal sea ice and snow cover is better known than the volume. Assuming average thicknesses of 1.5 m and 2 m for the ice cover that melts each summer in the Southern Hemisphere and Northern Hemisphere, respectively, the average (1979–2010) sea-ice contribution to this latent energy budget is approximately 11×10^{21} J per year (11 ZJ), equivalent to 349 TW (table 8.1). This neglects the energy required to warm the ice to the melting point. Compare this with a global energy consumption of 14.8 TW in 2009. This latent energy is drawn from ocean surface waters and the atmosphere.

The depth of the seasonal snowpack has a similar amount of spatial variability to sea-ice thickness, ranging from less than 100 mm w.e. in interior steppe and

Table 8.1
Estimated Latent Energy Required to Freeze/Melt the Seasonal Components of the Cryosphere Based on the Average Minimum and Maximum Sea-Ice Area (1979–2010) and Snow-Covered Area (1967–2010)

Component	Area $(10^6 \, km^2)$	H $(m \, w.e.)$	E_m $(10^{21} \, J \, yr^{-1})$	P_m $(10^{12} \, W)$
Seasonal snow	44.8	0.3	4.50	142
Northern Hemisphere sea ice	8.8	1.8	5.31	168
Southern Hemisphere sea ice	12.6	1.35	5.70	181
Active layer	22.8	0.18	1.37	44
Total	—	—	16.87	536

Note: H is an estimate of the average snow/ice thickness, E_m is the latent melt energy, and P_m is the average annual rate of transfer of latent energy, in TW.

tundra environments to more than 2000 mm w.e. in high-latitude coastal and mountain regions. Taking 300 mm w.e. as an estimate of the mean and using the area of seasonal snow cover from chapter 1, melting of the seasonal snowpack requires a total of 4.5×10^{21} J per year (142 TW). This is less than half of the energy consumed by melting of seasonal sea ice, but it is all derived from the atmospheric energy budget. This amount of energy is released to the troposphere each year through condensation of water vapor and freezing/deposition of snow crystals, and then consumed during melt.

Taking a similar approach for the seasonal freeze–thaw of the active layer in permafrost, with the assumption of an average active layer depth of 1 m, with 20%

ice content, active layer phase changes involved an additional 44 TW of latent energy. The total latent energy cycled in seasonal snow and ice is therefore about 536 TW. Additional latent energy exchanges are associated with seasonally frozen ground, lake ice, and river ice; these are difficult to estimate but likely contribute an additional 10–20 TW. This is an enormous amount of energy. Only 3% of the total solar energy available annually at the Earth surface is incident in the latitudes 60° to 90° (both hemispheres combined). This represents about 2400 TW in latitudes above 60°, where the majority of the snow/ice melt energy is consumed each summer. The latent energy sink therefore represents about 20% of the available solar radiative energy at these latitudes.

Note that for all of the seasonal cryosphere, the latent energy budget averages to near zero over a year (no net source or sink), depending on the state of the cryosphere. This foreshadows discussions of cryospheric change in chapter 9. Melting of glaciers and ice sheets over the past several decades has introduced an additional, unidirectional energy sink. The energy committed to this is consequential. For the period 2002–2009, for instance, the world's glaciers and ice sheets melted at an average rate of about 750 Gt per year. The energy required for this amounts to 8×10^{12} W (8 TW). Thinning Arctic sea ice and permafrost add to this cryospheric energy sink. These reductions in the cryosphere are acting as a thermal buffer that reduces the severity of atmospheric and oceanic warming. The efficacy of this buffer will diminish as the cryosphere contracts.

OCEAN–ICE INTERACTIONS

The melting and freezing of sea ice moderates ocean temperatures, as discussed in the preceding section. In addition, this affects the salinity budget in polar seas, with ramifications for ocean circulation and formation of water masses.

Sea-ice formation results in brine expulsion and the creation of dense, saline bottom waters. For the near-freezing temperatures of the polar ocean, density is primarily governed by salinity. Katabatically maintained polynya adjacent to the Antarctic coast are "sea-ice factories" associated with large rates of bottom-water production. This is particularly effective in wide parts of the Antarctic continental shelf in the Weddell Sea, where brine expulsion leads to formation of *high-salinity shelf water.* This mixes with the highly saline circumpolar deep water that sits just off of the shelf to create *Antarctic bottom water*, which floods much of the world's deep ocean abyss. Circulation and melting beneath ice shelves probably play some role in this, as ice shelf meltwater can be exceptionally cold, and it mixes with high-salinity shelf water to contribute to the deep water mass outflow from Antarctica.

In the Arctic Ocean, sea-ice formation has a similar effect, but saline deep waters are largely confined to the Arctic basin. However, sea-ice formation in the Labrador Sea and the Scandinavian seas contributes to intermediate and deep-water formation in the North Atlantic.

Sea-ice melting has the opposite effect, freshening surface waters where it melts. In the Arctic, this combines with high quantities of spring and summer river runoff to contribute to strong stratification of surface waters; surface waters in the Arctic are the freshest of any of the world's oceans. Glacial meltwater also contributes to this freshening, in particular runoff from the Greenland ice sheet. Annual meltwater discharge from Greenland is similar in magnitude to the major Arctic river basins that drain Canada and Russia, ca. 350 km^3 w.e. yr^{-1}, with most of this concentrated in the summer months.

Where water masses and sea ice advect out of the Arctic basin, through Fram Strait and the Canadian Arctic Archipelago, this represents a significant freshwater export, which can also tilt the salinity budget of the North Atlantic. Major freshwater advection events have been observed in the past, such as the "Great Salinity Anomaly" that began in the late 1960s. Strong ice export through Fram Strait in 1967 was identified as a pool of freshwater in southeast Greenland in 1968, which was then transported in the subpolar gyre into the Labrador Sea. It stalled there until 1972, before getting caught up in the North Atlantic Drift and advecting back into the northeastern Atlantic. The low-density surface water anomaly could be tracked until 1982. This was a long-lived feature that disrupted convective mixing and intermediate water formation in the Labrador Sea, creating cooler than usual sea-surface temperatures. This event was strong but does not appear to have been unique; several multiyear low- and high-salinity anomalies have been tracked in the North Atlantic region, linked with variations in freshwater export from the Arctic.

Icebergs that issue from the polar ice sheets, ice caps, and coastal tidewater glaciers cool and freshen the waters in which they melt. Icebergs can have keels that are hundreds of meters deep, and melting at depth gives freshwater plumes that promote mixing and ventilation. Algal blooms have been observed as a result of the nutrient delivery that accompanies this. In tidewater fjords and estuaries, most icebergs ground and melt locally, helping to stratify surface waters and strengthen the estuarine circulation. In embayments with shallow sills there is limited mixing with offshore waters, so strong stratification associated with freshwater runoff and iceberg melt can lead to isolation of a basin, suppressed ventilation, and anoxia at depth.

During the Pleistocene glaciations, large quantities of icebergs coming off of the North American and Eurasian ice sheets created a freshwater perturbation that was sufficient to disrupt intermittently the overturning circulation in the North Atlantic. Episodic iceberg fluxes from Hudson Strait, known as *Heinrich events*, flooded the North Atlantic region several times during the last glacial cycle, creating a stable cap of cold, fresh surface waters that helped to maintain the icebergs and enabled them to advect all the way from the Labrador Sea to the coast of Portugal. These events had worldwide climate effects, telegraphed through both the atmosphere and through disruptions of the North Atlantic deep-water formation. Meltwater runoff from the Pleistocene ice sheets and from catastrophic drainage of massive glacier-dammed lakes created similar disruptions to ocean stratification and circulation during the glacial period and in the early stages of deglaciation (8200 years ago).

In today's interglacial world, the tables appear to be turned, and the oceans have begun to perturb the polar ice sheets. Marine-based outlet glaciers and ice shelves in Greenland and Antarctica melt and calve off at the ice–ocean interface, with ablation at this interface exceptionally sensitive to ocean temperatures. This also occurs in many of the smaller ice caps in the Arctic, where ice extends to the sea. Ocean warming or wind-driven changes in ocean circulation that bring warm water masses (e.g. circumpolar deep water or North Atlantic water) in contact with the ice can trigger destabilization of ice shelves and marine-based outlets. This propagates inland through thinning, grounding-line retreat, and accelerated ice flow, in a classically known "tidewater glacier" instability.

Although the process is well understood for tidewater outlet glaciers in places like Alaska, it is not known how long it can continue and how dramatic it will prove to be for major marine-based sectors of the polar ice sheets. Rapid, ongoing changes in the Amundsen Sea sector of West Antarctica and Jakobshavn Isbrae in Greenland in the 2000s are linked with advection of warm-water anomalies to each region.

Melting of the Greenland and Antarctic ice sheets is unlikely to perturb large-scale ocean circulation, as ice sheet fluctuations did during the Pleistocene glaciations. The landscape was much different at that time, with permafrost, ice sheet lobes, and proglacial lake systems covering much of the midlatitude land mass in the Northern

Hemisphere. Greater energy is available for melting ice sheets at low latitudes, and the glaciated area contributing to runoff was large compared with that of today. The quantity of meltwater that characterized the glacial period is not conceivable from the Greenland and Antarctic ice sheets. Under most climate warming scenarios for the coming centuries, the freshwater fluxes associated with increased midlatitude precipitation, Arctic export, and changes in sea ice are greater in magnitude. Dynamical destabilization of a major sector of the Greenland or Antarctic ice sheet could provide an exception to this. Ice sheet processes that could deliver large freshwater fluxes to the ocean include mechanical disintegration, as observed in the Larsen Ice Shelf, or a major surge event, similar to what must have occurred during Heinrich events.

INFLUENCES ON ATMOSPHERIC CIRCULATION

In the winter months, the cooling influence of snow-covered surfaces helps to promote the creation of dense, high-pressure ("continental polar") air masses, which penetrate to midlatitudes as cold fronts. Strong winds and snowfall are followed by cold, clear weather and temperature inversions as continental polar air masses set up over a region. These air masses form over sea ice and over high-latitude (i.e. snow-covered) land masses during the winter months due to longwave radiative cooling, low amounts of incoming solar radiation, and the cold surface area.

In the Northern Hemisphere, cold fronts are associated with upper-air troughs and southward displacement of the polar front over the continents, driving Rossby wave structure and meridional mixing (baroclinic instabilities). The thermal contrast between snow-covered land and open ocean in the midlatitudes of the Northern Hemisphere promotes strong temperature gradients that help to drive this mixing. This results in a mobile polar front and alternation between cold, dry (boreal) and mild, wet (southerly or westerly) air masses in the winter months in much of North America, Europe, and Asia.

Similar air-mass cooling occurs year-round over Antarctica, driving the ferocious katabatic winds that descend from the continent. Unlike the Northern Hemisphere, however, high latitudes in the Southern Hemisphere have a relatively simple land–sea configuration, so circulation in the ocean and atmosphere is strongly zonal. The deep cooling influence of the high-elevation, perennially snow-covered continent helps to strengthen the Antarctic vortex, further isolating the continent from warm, midlatitude air masses.

Greenland also produces katabatic winds, and more modest glacier winds descend from most mountain glaciers and icefields. Some of these are true katabatic winds (gravity-driven, associated with cold, dense air masses created in the upper catchment), and some of these glacier winds are more of a result of topographic funneling of regional winds, particularly for valley glaciers. Regardless of their dynamical origin, glacier winds exert a strong local cooling influence in summer months, when the 0°C surface of a melting glacier cools the air in the glacier boundary layer. This cooling influence extends to both the glacier surface and the adjacent, downwind region.

Large ice sheets like those in Greenland and Antarctica also influence climate by presenting a topographic obstruction to tropospheric circulation. Similar to mountain belts, air flows around these obstacles, and much of the time they create persistent high-pressure (anti-cyclonic) circulation patterns. During the Pleistocene glaciations, ice sheets over North America and Eurasia disrupted the midlatitude westerlies and the prevailing pattern of stationary waves, leading to splitting of the polar jet stream over North America and increases in precipitation south of the ice sheets. Expanded sea-ice cover over the North Pacific and North Atlantic during the glaciation also modified the stationary wave pattern and the associated storm tracks, through southern displacement of the Aleutian and Icelandic Lows.

Because snow and ice affect pressure patterns, interannual variability in seasonal snow and ice may also play a role in patterns of variability in high-latitude atmospheric circulation, such as the North Atlantic Oscillation and the Arctic Oscillation. This is not fully understood.

CRYOSPHERIC EFFECTS ON THE HYDROLOGICAL CYCLE

Snowfall is an integral part of the global hydrological cycle. Even at low latitudes, much of the precipitation originates as snow crystals in the atmosphere. Some aspects of snow hydrology are discussed in chapter 4. With respect to the global hydrological cycle, one of the main effects of snow cover is to temporarily store water on the continents and on sea ice, delaying its return to the ocean by weeks to months. In the case of glaciers and ice sheets, water can be locked up on the continents for years to millennia. In snow melt–dominated hydrological catchments, peak river flows occur during the spring freshet, when runoff from snow melt reaches its maximum. Glacier meltwater supplements this in late summer and during the dry season(s) in the tropics.

Lakes and oceans affect regional climate in several ways. Open water in winter months provides a source of longwave and sensible heat fluxes, moderating the climate of coastal areas and in locations next to major lakes. This effect vanishes once sea or lake ice takes hold for the winter. Open water also provides a source of moisture, increasing the snowfall in downwind areas. This effect is well documented in the Great Lakes area of North America, in coastal ice caps adjacent to the North Water polynya in Baffin Bay, and in southeastern Greenland.

Warmer temperatures and loss of sea ice are expected to increase atmospheric moisture and snowfall at high latitudes. This creates a climate-change feedback that makes the system unpredictable. Increased snowfall has a positive influence on glacier mass balance, a negative impact on permafrost thickness, and mixed effects on lake and sea-ice thickness, depending on the seasonality of snow cover. A deeper snowpack in the winter insulates the ice and limits its growth, giving a thinner ice pack.

The formation of snow ice, through snowpack loading and submergence, has opposite effects on ice thickness. Spring and summer snowfall on lake ice, sea ice, and glaciers also promotes thicker ice, as the increased albedo helps to limit melting.

CRYOSPHERE–BIOSPHERE INTERACTIONS

There are numerous interesting ecological influences of the cryosphere, such as marine mammals' penchant for the sea-ice edge, the hibernation rhythms of bears, the deadly nature of rain-on-snow events for Peary caribou, and the importance of deep snowpacks for winter feeding of woodland caribou, which, lacking giraffe necks, rely on the extra elevation to forage higher into spruce, fir, and cedar trees. Photosynthetic activity and ecological rhythms of high-latitude rivers and lakes are also closely connected with the ice cover.

Snow, ice, and permafrost play an interesting but poorly understood part in the global carbon cycle. Polynya tend to be nutrient-rich, ecological hot spots, associated with CO_2 drawdown into the ocean, so it is suspected that open water (versus sea-ice cover) would be conducive to a stronger sink for atmospheric carbon in the polar regions. There are algae that thrive in sea ice, however, so carbon–ice–ocean exchanges are not fully predictable.

Glacier and permafrost cover alter carbon storage in the landscape. Glaciers and ice sheets override vegetation and soil, effectively removing this carbon from the system. Some of this organic carbon can be stored sub-glacially, as evidenced by retreating glaciers, but most of it decomposes and is washed out via subglacial meltwater, leaving a relatively barren, inorganic subglacial and periglacial environment. Soil is still trying to establish itself in the till deposits of most glacier forefields, which have been exposed by glacier retreat since the sun set on the Little Ice Age in the late 19th century.

Permafrost in much of northern Canada, Russia, Alaska, and Scandinavia is found in organic-rich mus-keg and peatlands, where there is high soil moisture and carbon content. When frozen, this carbon is removed from the atmospheric cycle, and it can be locked up in the ground for long periods. As introduced in chapter 7, thawing of permafrost and deepening of the active layer in recent decades are causing a reverse effect: the release of soil carbon to the atmosphere, as both methane (CH_4) and carbon dioxide (CO_2). Carbon fluxes from thawing permafrost may be offset by increased carbon uptake as vegetation and biomass expands at high elevations and in high northern latitudes. Similarly, biological carbon uptake may increase in high-latitude ocean waters and as the snow-free growing season is extended in midlatitudes. The net impact of cryospheric change on the carbon cycle is therefore unclear.

Variations of CO_2 and CH_4 during glacial–interglacial cycles pose an ongoing puzzle to our understanding of the carbon cycle. Levels of both greenhouse gases dropped systematically during the Pleistocene glaciations, as documented by the air bubbles trapped in glacial ice in Greenland and Antarctica. Ice cores show a remarkably close correlation between the concentrations of CO_2 and CH_4, air temperature, and the volume of ice on the planet, so it is clear that greenhouse gas reductions acted as an important feedback in driving the world in and out of glaciations. However, the nature of the glacial carbon sink remains uncertain. The quantity of carbon in the ice sheets is negligible, so the carbon sink lies elsewhere.

The direct effect of the Pleistocene ice sheets should be to release carbon from overridden vegetation and soil to the atmosphere, decreasing terrestrial CO_2 and increasing atmospheric levels. This means that even more CO_2 must have been taken in by the oceans or by tropical vegetation during the glaciations. One aspect of the cryosphere may be implicated in the glacial carbon sink: carbon sequestration in frozen ground. Midlatitude permafrost may have acted as a major carbon sink during the glaciation, both under the ice sheets and in the proglacial regions. If so, the coupling between ice sheet advance/retreat and uptake/release of this carbon was extremely tight, implying a century-scale response time for permafrost carbon storage.

SUMMARY

The cryosphere is an integral part of the climate system, shaping the weather, climate, and many aspects of society in midlatitudes and high latitudes. Cryospheric processes and feedbacks also affect the global climate through influences on the planetary energy budget and circulation in the oceans and atmosphere. Indeed, the cryosphere has played a lead role in some of the greatest climate perturbations in Earth's history. It is also reacting strongly to recent climate change. Some aspects of climate–cryosphere variability are discussed in the next chapter, building on the cryosphere–climate processes outlined in this chapter.

9 THE CRYOSPHERE AND CLIMATE CHANGE

....

All is on the eve of motion.
—James Forbes,
Travels Through the Alps of Savoy

SNOW AND ICE TAKE ON MANY GUISES ON EARTH, SOME fleeting and some older than civilization. Cryospheric dynamics have played a role in climate evolution throughout Earth's history. Several examples of cryosphere–climate interactions were introduced in chapter 8. Here, I present additional examples of cryospheric change and the influence of the cryosphere in the climate system, including a brief look at "snowball Earth," the evolution of the Antarctic and Greenland ice sheets, a fuller discussion of glacial–interglacial cycles, and a brief overview of recent cryospheric change. The planet is changing rapidly, so some discussion is warranted. However, this area represents a moving scientific target, so I restrict the discussion to a general summary rather than a detailed analysis of current climate and cryospheric trends.

....

THE CRYOSPHERE IN THE DISTANT PAST

For decades, Earth scientists argued that the continuous persistence and success of life on Earth for more than 3 billion years is clear proof of a Goldilocks state: never too hot, never too cold for life to persevere. Earth somehow managed to avoid extreme states such as a runaway greenhouse effect where the oceans boiled away or, more germane to this chapter, a "snowball" state where the planet was completely ice-covered. The latter would be not only tough on life but also an extremely difficult state from which to escape. The positive cryospheric feedbacks discussed in chapter 8, in particular the effect of ice-covered tropical regions on the global albedo, would limit the available energy that could be harnessed to melt away the ice and break out of the snowball state.

This view has been challenged by an accumulating body of evidence indicating global or near-global ice cover on at least two occasions in the deep past: the Sturtian glaciation, ca. 720 Ma (million years before present), and the Marinoan glaciation, ca. 640 Ma. There is evidence from tillites (ancient glacial tills) and ocean sediments for tropical ice cover at this time. The Earth may have escaped an ice-covered state through buildup of CO_2 in the atmosphere as a result of continued volcanic outgasing, combined with the absence of terrestrial and oceanic carbon sinks. Evidence from cap carbonates in the ocean supports this hypothesis. The meltdown process would have been aided by the gradual darkening (decreased albedo) of snow and ice as open-water moisture sources became cut off and fresh snowfall became limited in the hyperarid, snowball world. The climate conditions that initiated snowball Earth events are uncertain. The Sun was about 6% weaker during the Sturtian,[1] which would have aided the process, but this does not explain what triggered the Neoproterozoic events.

The snowball Earth hypothesis remains controversial, and new ideas, geological evidence, and modeling studies are fueling ongoing debate. There are arguments for only partial ice cover during these two major events, with the tropical oceans remaining open. Others argue for additional global glacial events throughout Earth's history. Although the story is still emerging, it is clear that major glaciations affected early Earth, probably assisted by the weaker Sun at this time, and there have been times when the cryosphere overwhelmed the entire planet for millions of years. Earth's oscillation between ice-covered and ice-free states is indicative of a delicately balanced climate. This climate sensitivity is largely a result of Earth's equilibrium temperature being so close to the triple point of water, along with cryosphere–climate feedbacks that tend to amplify perturbations to our mean state.

BIRTH OF THE POLAR ICE SHEETS

Excepting the massive blips associated with episodes of global or near-global ice cover, the Earth has been ice-free for much of its history. During the Paleozoic and

Mesozoic eras, from about 550 to 65 Ma, life expanded dramatically on the planet, and there is little evidence of glaciations through this time. The Cretaceous period, from 144 to 65 Ma, was the apex of the Mesozoic, characterized by sea levels 200 m higher than present, deep ocean temperatures of up to 15°C, and tropical flora and fauna in the polar regions. The cryosphere most certainly had no presence at this time, with the likely exception of occasional snowfalls in polar regions and on the world's highest mountains.

The Cenozoic era, which followed the Cretaceous, marked a turnaround for planetary temperatures and the cryosphere. The Earth began to cool at about 50 Ma (figure 9.1), most likely driven by reduced carbon dioxide levels in the atmosphere and associated water vapor feedbacks. Reduced tectonic outgasing and increased weathering sinks for CO_2 as a result of major mountain-building events have been invoked to explain this, although there are still many questions as to what has driven Earth's long-term climate evolution. Increased photosynthetic activity and carbon sequestration as the global oceans cooled probably played a role in Cenozoic climate change. Cryosphere–climate feedbacks also had an influence, through the slow return of snow and ice to the landscape and the birth of the Antarctic and Greenland ice sheets.

Antarctica began to accumulate ice by 35 Ma, most likely through glacial inception from mountain glaciers and icefields that spread to lower elevations as climate cooled. By 23 Ma, the Drake Passage and Tasmanian

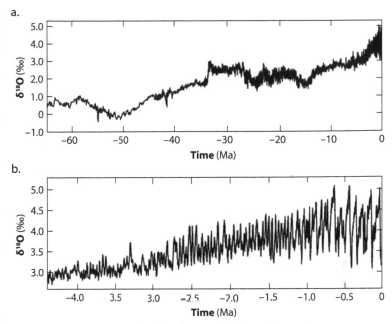

Figure 9.1. (a) Marine benthic $\delta^{18}O$ records from the Southern Ocean for the past 65 million years, chronicling the Cenozoic cooling and several major cryospheric events, including the period of Antarctic ice sheet growth ca. 15 Ma, and the onset of the Quaternary glacial cycles at about 2.5 Ma (Zachos et al., 2008). (b) Zoom-in on data for the past 4.4 Ma (Lisiecki and Raymo, 2005). Higher values of $\delta^{18}O$ in marine sediments indicate both cooler temperatures and greater ice volume on the continents.

Passage were opened, as divergent motion of the tectonic plates separating Antarctica from South America and Australia progressed to allow a continuous oceanographic passage. This allowed the Antarctic Circumpolar Current to develop, largely isolating Antarctica from midlatitude air and water masses. Ocean sediments document the transition that took place in Antarctica at this time, as ice displaced coniferous forests and something resembling the modern-day ice sheet took root on the continent. East Antarctica is believed to have been continuously ice-covered since about 15 Ma. The West Antarctic ice sheet may have developed more recently, during the past 10 million years (Myr). Modeling studies indicate that the extent of West Antarctic ice probably fluctuated during the past 5 Myr, partially collapsing and regrowing in response to cyclical variations in Earth's axial tilt (the same orbital forcing that drives Northern Hemisphere glacial cycles).

Once established, the Antarctic ice sheet grew to a thickness of several kilometers, further cooling its own climate, strengthening the Antarctic vortex, and promoting cold air outflows that help to cool the Southern Ocean and drive increased Southern Hemisphere sea-ice extent. The associated albedo feedbacks contributed to large-scale Cenozoic cooling. Expanded sea-ice production and colder surface ocean waters are necessary ingredients for the formation of Antarctic bottom water, which circulates through most of the world's deep ocean. The growth of the Antarctic ice sheet was therefore a major climate event in Earth's recent history.

As global cooling continued through the late Cenozoic, the Arctic also began to support seasonal snow and sea ice during the past several million years. Ice-rafted debris indicates that mountain glaciers and icefields in eastern Greenland must have flowed down to the coast at times during the past 10 Myr, and by 3 million years ago this ice cover had expanded to most of the island. The Greenland ice sheet has waxed and waned with the Quaternary glacial cycles of the past 2.5 Myr, but a central core of ice may have persisted throughout this time.

The closure of the Panama seaway and the development of the Isthmus of Panama isolated the North Atlantic basin between 4 and 5 Ma. It has been argued that this increased the moisture transport to high northern latitudes, promoting the development of the Greenland ice sheet and the Quaternary glacial cycles in the Northern Hemisphere. This undoubtedly had an impact on the North Atlantic gyre, and the North Atlantic deep-water formation that we know today may have had its onset at this time. It is unclear, however, whether this had a major role to play in Northern Hemisphere glaciations. The timing is too early, and substantial high-latitude warming accompanied the increased moisture, which generally has a stronger influence than precipitation on the viability of ice sheets. It is more likely that the cooling trend of the past 5 Myr allowed temperatures to transgress a threshold, beyond which it became cold enough to allow the Greenland ice sheet to expand and develop a permanent presence. Ice–climate feedbacks would have supported its survival once established. The flickering ice sheets in North America and Europe soon followed.

PLEISTOCENE GLACIAL CYCLES

The world changed again beginning at about 2.5 Ma. Figure 9.1b illustrates the onset of the Pleistocene glacial cycles at this time, a turbulent period in Earth history that continues to this day. The planet has experienced more than 40 glacial–interglacial episodes during this period, with ice sheets intermittently flooding large expanses of the Northern Hemisphere continents. As seen in Figure 9.1b, these cycles increased in magnitude over the past million years, indicating that glaciations have become more severe. There has also been a change in the periodicity, with the *mid-Pleistocene transition* at about 1.2 to 0.9 Ma marking a shift from 40-kyr to 100-kyr glacial cycles. This is approximate; the duration of individual glacial and interglacial episodes varies.

The last glacial cycle began about 116,000 years ago (116 ka), with the inception of ice sheets in northern Canada and Scandinavia. Ice spread fitfully over North America, Europe, and parts of Asia, South America, and New Zealand for about 100,000 years, while the Greenland and Antarctic ice sheets also grew, expanding out to the edge of the continental shelf. The ice sheets reached their maximum extent 21,000 years ago, a moment known as the *last glacial maximum*. The ice sheets began to recede at this time, largely disappearing from the landscape by 8 ka, with the notable exceptions of Greenland and Antarctica. Vestiges of the Laurentide ice sheet also remain in the Canadian Arctic Archipelago.

The last glacial cycle is commonly called the "Ice Age," although this overlooks the fact that Earth has experienced numerous, repeated glaciations. We have not seen the last of the great ice sheets, although we are currently enjoying an interglacial respite. Barring too much human interference with the climate system, the current interglacial period is certain to come to a close, perhaps 20,000 years from now. Wallace Stegner captures this nicely in his musings about the Canadian climate:

> The ice sheet that left its tracks all over the region has not gone for good, but only withdrawn. Something in the air, even in August, says it will be back.

The Quaternary glacial–interglacial cycles personify cryospheric sensitivity to climate change and the dynamic, central role that the cryosphere can play in Earth's climate. Although continental ice sheets leave the strongest imprint on the landscape, and indeed in the ocean, ice sheet advance during glacial periods was accompanied by large-scale expansions of all aspects of the cryosphere; snow, freshwater ice, permafrost, and sea ice all extended to lower latitudes and contributed to the cryosphere–climate feedbacks that supported both the advance and retreat of the ice sheets, once under way.

Textbooks have been dedicated to discussions of the Pleistocene glaciations. Here, I provide a brief sketch of some important aspects of glacial cycles, with an emphasis on cryosphere–climate feedbacks. Michael Bender's text *Paleoclimate* examines glacial cycles in greater detail.

Orbital Variations

High-latitude orbital forcing has been demonstrated to pace the glacial–interglacial cycles. This is not a matter of changes in solar output, but of geographic and seasonal variations in the distribution of insolation over the planet. Gravitational influences from the Moon and other planets (primarily Jupiter and Saturn) introduce systematic, cyclic variations in the Earth–Sun orbit, including changes in (i) the eccentricity of Earth's orbital path around the Sun, (ii) Earth's axial tilt angle, relative to the normal to the plane of the ecliptic, also known as the *obliquity*, and (iii) the direction of Earth's tilt angle, relative to the celestial sphere. The latter effect, also known as *precession*, matters because it determines the season when Earth marks its closest approach to the Sun, a time known as perihelion. Orbital conditions change on time scales of 10^4 to 10^5 years: specifically, ~100- and 413-kyr cycles for eccentricity, 41-kyr cycles for the tilt angle, and ~19- and 23-kyr cycles for precession.

Perihelion currently falls on January 3, which means that Southern Hemisphere summers are more intense than Northern Hemisphere summers. The current value of our orbit's eccentricity is 0.0167, and this ranged from 0.005 (a nearly circular orbit) to 0.058 during the Pleistocene. This has only minor effects on global annual insolation, but the seasonal effect is substantial. Earth currently receives almost 7% more solar radiation at perihelion than aphelion (1415 vs. 1323 W m^{-2}), and the difference increases to 23% when the eccentricity is at its most extreme value. Earth's axial tilt, currently 23.54°, oscillates from 22.1° to 24.5°, further modifying seasonality at high latitudes.

Orbital variations are also known as *Milankovitch cycles*, named for the Serbian mathematician who calculated the effect of these variations on the seasonal radiation at high northern latitudes. Milankovitch spent many years making the necessary calculations to build on James Croll's hypothesis that glaciations are a result of changes in seasonal insolation in the Northern Hemisphere; cold summers are needed for seasonal snow to survive and transform to perennial snowfields, which eventually grow large enough to flood the landscape. The Northern Hemisphere is the key as this is where the high-latitude land masses reside, and this is where the former ice sheets made their mark.

Milankovitch did not live to see the confirmation of his theory. His *Canon of Insolation of the Earth and Its Application to the Problem of the Ice Ages*, published in 1941, was met with skepticism and was not translated to English until 1969. By then, deep-sea sedimentary records, understanding of oxygen isotope records in calcite shells from marine sediments, acceptance of plate tectonics, and advances in geochronology and time series analysis (e.g., Fourier transforms) all came together to make it possible to examine quantitatively and interpret the climatic history documented in ocean sediments. The periodicities of the orbital variations are strongly echoed in the marine $\delta^{18}O$ record of glacial–interglacial cycles. Orbital variations are now well established as the "pacemaker of the ice ages"; ice sheets expand over the northern continents when eccentricity, tilt, and precession align to give cool summers at high latitudes, and the ice retreats when these elements are oppositely aligned, giving warm northern summers.

Internal Climate System Feedbacks in Glacial Cycles

The link between the periodicity and phasing of orbital variations and the Quaternary glacial cycles is convincing, so there is little doubt that orbital variations play a central role. A great deal is unexplained, however. Much of the original skepticism over the orbital theory stems from the fact that the perturbations to insolation are relatively modest and are hard to reconcile with the magnitude of the climate system response. Sophisticated climate models have been brought to bear on the problem, with little success in simulating either the glacial inception or the deglaciation. This may be a matter of resolution and timescales; it is difficult to conduct millennial-scale integrations with general circulation models (GCMs), the highland locations of ice inception centers are not well resolved, and glacier dynamics (absent in most GCMs) are needed to include the effects of ice advection from inception centers to lower altitudes and latitudes. Long integrations including ice-albedo feedbacks and other climate system feedbacks are needed to describe this process.

The current view is that orbital forcing acts as the trigger for glaciation and deglaciation, but most of the climate forcing that drives glacial–interglacial cycles in fact comes from feedback mechanisms and accompanying shifts in global climate dynamics. Many of the relevant processes are cryospheric, such as the collective albedo and regional cooling impact of expanded sea ice and snow cover, which adds to the area of ice sheets. In addition, increased midlatitude freshwater delivery to the oceans, via iceberg calving and summer melting on the ice sheet margins, affected ocean circulation in the North Atlantic region, reducing deep-water formation and poleward heat transport. The orographic disturbance of the ice sheets also altered patterns of atmospheric circulation.

As discussed in chapter 8, greenhouse gas levels also dropped dramatically during the glaciations. Atmospheric water vapor decreased as the planet cooled, and ice cores indicate that carbon dioxide and methane levels were much lower than preindustrial values. The correlation between Antarctic temperatures, global ice volume, and concentrations of CO_2 and CH_4 hold true over the past several glacial cycles (900 kyr), as far back as the ice cores reach. While the causal link is still enigmatic, greenhouse gas reductions played a crucial role in helping to cool the planet in glacial times, and increases in CO_2 and CH_4 also helped to drive the ice sheet demise during each deglaciation. Changes in oceanic carbon uptake associated with colder oceans, altered ocean circulation, and changes in ocean alkalinity during glaciations are the most likely explanations, although shifts in terrestrial carbon uptake (e.g., methane storage in tropical wetlands, midlatitude permafrost, and subglacial environments) may have played a role.

Overall, glacial cycles provide the textbook example of nonlinear feedback mechanisms in Earth's climate dynamics, which are able to take a relatively small orbital forcing—a seasonal and geographic redistribution of insolation—and amplify it into a global climate shift that transforms the landscape. Cryospheric feedbacks and ice–atmosphere–ocean processes were central to this.

Another aspect of the Pleistocene glacial cycles is important to note in the context of future climate change. There is palynological, sedimentary, ice-core, and sea level evidence that some previous interglacial periods were warmer than the present, in particular isotope stages 11 (ca. 400 ka) and 5e (the Eemian, ca. 125 ka). Arctic sea ice declined at these times, and the southern sector of the Greenland ice sheet retreated dramatically in both of these periods, contributing to sea-level high stands that were 5 to 6 m higher than present. Proxies and modeling indicate that the Arctic was up to 5°C warmer than present in the Eemian, driven by high spring and summer insolation. These periods may be good analogues for the Arctic environment in a warmer world.

Millennial Climate Variability During Glacial Periods

Ice core records and high-resolution marine sedimentary records indicate that glaciations are also associated with "suborbital" climate fluctuations, with dramatic variability on timescales of centuries to millennia. Many of these events likely involve ice sheet dynamics and other cryosphere–climate instabilities, such as regime shifts in sea ice or freshwater forcing of the oceans. The best-documented examples of millennial climate variability are Dansgaard–Oeschger cycles and Heinrich events.

Dansgaard–Oeschger (D-O) cycles are ~1500-year climate oscillations that characterize the glacial period, with their strongest expression in Greenland ice cores. In Greenland they are marked by abrupt (decadal-scale) temperature oscillations of as much as 15°C, with concomitant changes in dust transport and snow accumulation rates. Snow accumulation rates in central Greenland more than doubled during the warm phase of D-O cycles. These events are centered in the North Atlantic region, but there are indications of effectively instantaneous, global teleconnections to the North Atlantic climate variability. D-O cycles are associated with and potentially attributable to fluctuations in meridional overturning circulation in the North Atlantic. Reduced deep-water formation during the cold phase of D-O events likely caused reductions in poleward heat transport along with expansion of sea ice in the North Atlantic. Sea ice expansion in turn provided a positive feedback that contributed to the extreme temperature fluctuations and cold, dry conditions in Greenland, through southward displacement of the North Atlantic Drift, the polar front, and the associated baroclinic activity (hence heat and moisture transport).

The role of the continental ice sheets in D-O cycles is unclear. It has been argued that high quantities of ice sheet meltwater or perturbations to the locations where this water reached the ocean weakened the meridional overturning circulation or freshened North Atlantic waters to the point where the overturning strength was susceptible to stochastic fluctuations. Alternatively, D-O cycles may represent internal variability in the ocean or the sea-ice–ocean system or a modulation of external (e.g., solar) forcing. Marine margins of the ice sheets in North America, Europe, and Iceland responded to D-O climate variability through ice-marginal advance and retreat, with resulting changes in iceberg discharge, but this was likely a response to D-O climate fluctuations rather than a causal mechanism.

Heinrich events are a different matter. These intermittent, large-scale fluxes of ice from the Laurentide ice sheet left their mark in the North Atlantic through the deposition of enormous quantities of ice-rafted debris, cooling and freshening of North Atlantic surface waters, southward deflection of the polar front, and disruption of the meridional overturning circulation. Hudson Strait has been identified as the source region for the icebergs. Five or six Heinrich events occurred during the past glaciation, with events lasting from decades to a few centuries. Heinrich events recurred every 5 to 10 kyr once the Laurentide ice sheet was well-established over Hudson Bay and eastern North America, but they were not periodic. These events are attributed to surging or tidewater retreat of an ice stream in Hudson Strait, with the ice sheet disgorging a volume of ice equivalent to an estimated 1 to 3 msl. There was disruption of deep-water formation during these events, due to freshening of the North Atlantic associated with iceberg melting and reduced influxes of subtropical (i.e., Gulf Stream) water.

Surging and tidewater retreat are classical, well-documented examples of internal dynamical instabilities in present-day glaciers, although (thankfully) a collapse of the magnitude of Heinrich events has not been observed on the scale of an ice sheet. It is still an open question whether Heinrich events arose purely from internal dynamics or may have been triggered through a climatic perturbation. The former is more likely, because there is no obvious atmospheric trigger, but a more subtle climate forcing such as warmer ocean waters may have destabilized the ice stream. These events are important to understand, as similar behavior in the Greenland or Antarctic ice sheets would pose a large disruption to society.

RECENT AND FUTURE CRYOSPHERIC CHANGE

In more recent Earth history, temperature fluctuations of 1°C on interannual to centennial timescales have significant impacts on mountain glaciers, sea ice, river and lake ice, seasonal snow cover, and active layer depths. Fluctuations in precipitation and other meteorological variables are probably secondary to the influence of temperature, overall, although they can be significant on local and regional scales.

Cryospheric sensitivity to temperature changes is apparent through cold and warm phases of the Holocene (the last 10,000 years). Holocene temperature fluctuations of order 1°C, associated with solar and volcanic variability, have driven several periods of glacier advance and retreat in mountain regions. Direct observations and proxy records from the Little Ice Age, which persisted from about 1500 to the late 1800s, offer some of the best evidence for this, with a global expansion of sea ice, snow, and glacier cover through this period. This is nicely documented in literary, artistic, and cultural records from Europe and Iceland. Geological reconstructions from terminal moraines, lake sediments, dendrochronology, and cosmogenic dating evidence indicate that the maximum glacier advances of the Holocene occurred in this period throughout the Northern Hemisphere. Southern Hemisphere glaciers also advanced at this time, but in some locations (e.g., New Zealand) there are records of an even more extended middle Holocene advance.

The glacier advances of the Little Ice Age were the culmination of several millennia of "neoglacial" glacier advances following the early Holocene climate optimum, ca. 6 ka, when global glacier coverage was retracted relative to present day—Northern Hemisphere mountain regions are believed to have been largely bereft of ice during the climate optimum. The Holocene evolution of glaciers is broadly consistent with orbitally driven changes in summer insolation, as discussed in the earlier section "Pleistocene Glacial Cycles," with shorter-term climatic influences superimposed on millennial-scale orbital trends.

Since the late 1800s, other climatic influences—increase in solar output, relatively quiet volcanic activity, and the buildup of atmospheric greenhouse gases—have overridden orbitally driven cooling trends, driving recent warming and a strong cryospheric response. The cryosphere response is particularly marked in the

Northern Hemisphere, where climate warming has been greater. The most dramatic changes are at lower elevations and latitudes, where snow and ice exist close to their 0°C threshold for viability. This has an interesting consequence. The Arctic has warmed more than most other regions on Earth during the past 40 years, but some of the largest changes in the cryosphere are being felt at midlatitudes, where permafrost, river and lake ice, mountain glaciers, and seasonal snow cover are marginal. Changes are also dramatic in polar latitudes in the summer months, in particular for the Arctic sea ice and the ablation zones of the Greenland ice sheet and Arctic ice caps. In the northern winter and in most of Antarctica, temperatures are still so low that climate warming is not strongly affecting the cryosphere.

Although temperature changes are less important in cold environments, the highest elevations and latitudes may be experiencing increases in snowfall in both hemispheres, due to increases in atmospheric humidity and more frequent advection of midlatitude air masses into subpolar and polar latitudes, at least in the Northern Hemisphere. This is not enough to prevent a negative mass balance for most of the world's glaciers, but it is a feedback that partially offsets increases in melt. Increases in high-latitude snow also affect the hydrological cycle and the thermodynamics of sea ice, lake ice, river ice, and permafrost.

The sensitivity of the global cryosphere to small changes in temperature has proved invaluable in helping to understand and document climate variability in remote, uninstrumented parts of the planet. It also provides evidence that measured 20th century warming is not an artifact of urban heat island effects or local land-use changes in the populated environments where most direct, long-term observations are derived; significant cryospheric changes are under way in all parts of the world, most of these far removed from urban centers.

Changes in Seasonal Snow

During the past century, snowfall has been well monitored in populated regions of Europe, Russia, and North America. Much of this information is archived as snow depth or fresh snowfall totals, from which snow–water equivalence (SWE) must be estimated. Beginning in the 1960s, visible-wavelength satellite imagery has provided detailed mapping of snow-covered area in the Northern Hemisphere (see chapter 1).

Because snow extends to low elevations and latitudes, snow cover is very sensitive to warming or cooling. This is clear in the satellite record, which shows major declines in snow-covered area in spring and summer in the Northern Hemisphere. Spring melt shifted earlier by more than 2 weeks since the early 1970s, and the time of peak snow-covered area has shifted from February to January. The largest snowpack decline has occurred in midlatitude regions, roughly corresponding with the March–April 0°C to 5°C isotherms in the band 40° N to 60° N. The rate of change of Northern Hemisphere snow cover in these months is -0.62×10^6 km^2 per decade

(1967–2010). Summer (June through August) snowpack declined at -0.79×10^6 km^2 per decade from 1972 to 2010. This is equivalent to snow cover reductions of 7.5% and 52% in the spring and summer, respectively, relative to the mean values over the period of record. Winter snowpack is more robust, as temperatures remain below freezing from November to February in most areas of the Northern Hemisphere that experience seasonal snowfall. Annual average snow cover declined at a rate of -0.34×10^6 km^2 per decade from 1972 to 2010: a net loss of 5.2% for this period.

Observed reductions in snow cover are not spatially uniform. Snowfall has had no statistically significant changes in some regions, and others have seen increases in snow cover. Southern Canada (below 60° N) experienced an average of 78 snow days per year (days with solid precipitation) for the period 1955–2004, but station data indicate opposite trends in western and eastern parts of the country. Snow events in western Canada declined by 2.9 days per decade during this period, whereas central and eastern Canada experienced an increase of 1.7 snow days per decade: respective changes of −14.5 and +8.5 days over the 50-year period. The frequency of snow days does not provide direct information on total snowfall, as SWE may change with little or no shift in snow-covered area or snow frequency. This does not seem to be the general case in midlatitudes, however, as snow cover, frequency, and depth are declining in conjunction, whereas winter rainfall events are becoming more frequent.

There are limited direct observations of snowpack trends at high elevations. Terrain variability in alpine environments prohibits straightforward satellite assessments of snowpack. Snow transect and snow-pillow data in the western United States and in the European Alps indicate a general decline in snowpack at all elevations in the latter half of the 20th century, with local exceptions. Declining glaciers and earlier peak runoff in stream flow from mountain headwater regions are consistent with diminishing snow accumulations at high altitudes, but these observations could also be a consequence of warmer temperatures in spring and summer.

The record from the Southern Hemisphere is also sparse. Snowpack measurements dating from the early 1960s in the mountainous regions of southeastern Australia indicate similar trends to northern midlatitudes: no major changes in maximum winter snowpack, but significant declines in snow cover and SWE in late winter and spring. The signal from the mountain snowpack in New Zealand and the Andes is not as clear. Interannual variability is high in these regions, governed by South Pacific synoptic variability (e.g., ENSO cycles), potentially masking decadal-scale trends.

Changes in Lake, River, and Sea Ice

There is a detailed historical record of freeze-up and breakup of lake, river, and sea ice in many communities, particularly where ship navigation and ice roads are dependent on ice conditions. These observations are local and they often reflect near-shore conditions, so they may not be representative of regional or synoptic conditions. These are nevertheless valuable in documenting changes in ice cover during the past century, particularly for river and lake ice. In many freshwater ice settings, local observations are simpler to interpret, as water masses are limited in extent, there is less multiyear memory, and interannual variability in winds, pressure patterns, and currents have a minor influence compared with their impact on sea-ice conditions. Available river and lake records that are at least 150 years in length in the Northern Hemisphere point to shorter winters, with freeze-up occurring 8.7 ± 2.4 days later and breakup moving up by 9.8 ± 1.8 days over this 150-year period.

Over a shorter time frame, the last 30 to 70 years, regional compilations of the freshwater ice season indicate a large degree of variability. For recent decades, visible and microwave satellite imagery offer good discrimination between open water and ice conditions. Regional variations in the ice season are broadly consistent with temperature changes, with observations pointing to a shortening of the ice season by roughly 10 days per °C of warming. Changes in the snowpack can cause deviations from this relationship with temperature.

Microwave remote-sensing observations also provide a detailed view of changes in sea-ice cover since the late 1970s. Northern Hemisphere sea-ice cover has declined over this time in all seasons, with the strongest changes in late summer months. The March, annual, and September trends from 1979 to 2010 are −2.8%, −4.4%,

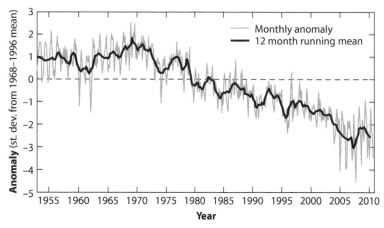

Figure 9.2. Monthly sea-ice extent anomalies, January 1953 to September 2010. Data from 1953 to 1979 is derived from operational ice charts and other sources (U.K. Hadley Centre). Passive microwave satellite data is available from 1979 to 2010. (Figure from W. Meier and J. Stroeve, National Snow and Ice Data Center, University of Colorado, Boulder, CO.)

and −12.4% per decade, respectively. In contrast, Antarctic sea ice has been stable over this period, with annual sea-ice extent in the Southern Hemisphere increasing by 1.3% per decade. This is consistent with observations of Antarctic cooling since the 1970s, attributed to the combined effects of stratospheric ozone depletion and strengthening of the Antarctic polar vortex, which helps to isolate Antarctica from meridional heat and moisture transport. Working from ice charts and ship records, records of sea-ice extent have been extended back beyond the satellite era, although with greater uncertainty. Figure 9.2 plots an example for the period 1953–2010 in the

Arctic, expressed as anomalies from the monthly mean ice extent. This illustrates the decline that began in the early 1970s. There are fewer observations available in the Southern Hemisphere, but whaling vessel records and sea-salt indicators in Antarctic ice cores indicate that Antarctic sea-ice cover was more extensive in the 1950s than today. This is consistent with the possibility that longer-term declines in Antarctic sea ice have been arrested since the 1970s by regional cooling associated with ozone depletion.

Ice thickness is more difficult to gauge than ice extent. Available evidence from upward-looking sonar imaging along submarine tracks indicates that the Arctic ice pack thinned by as much as 1.3 m, on average, from the 1950s to the 1990s. Satellite laser altimetry measurements of sea-ice draft have enabled estimates of ice thickness since the early 2000s (e.g., figure 5.2a). These records indicate that multiyear ice over the Arctic basin thinned from 3.6 to 1.9 m from 1980 to the late 2000s, but longer altimetry records are needed to assess what is typical for interannual and interdecadal variability in ice thickness. Basin-scale variations in ice thickness and extent should be broadly correlated, as both respond to changes in oceanic and atmospheric temperatures. More open water (reduced ice extent) leads to warming of the ocean mixed layer and increased ice melt, hence thinner ice. This general correlation can break down on regional scales, as ice convergence (e.g., due to persistent onshore winds) can give compaction and ridging, leading to reduced ice extent but unusually thick, concentrated ice.

Glacier and Ice Sheet Response to Climate Change

Glacier mass balance was introduced in chapter 6. A concerted, global effort to monitor glacier mass balance took root in the 1960s. Annual mass balance data have been gathered for a suite of at least 60 glaciers since 1964 and for at least 90 glaciers since 1981, peaking at 109 glaciers in 1998. Different glaciers have been monitored over this period, with more than 330 glaciers with at least 1 year of data. This is a small fraction of the estimated global population of more than 200,000 glaciers, so most of what is known about glacier mass balance is based on a small subset of glaciers. This is difficult to improve on due to the field-intensive effort required for classical mass balance studies.

Extrapolating from the global set of mass balance data and the areal distribution of glaciers, regional, continental, and global glacier changes can be estimated from the early 1960s to the present. Figure 9.3 plots an example of this for the global distribution of small glaciers and ice caps, excluding the polar ice sheets.

Interannual variability in figure 9.3 is muted relative to individual mass balance time series, due to averaging from different regions that experience asynchronous climate excursions. A number of global climate trends and events are still evident, such as the cooling that followed the 1991 eruption of Mt. Pinatubo. This produced the only year with a positive net mass balance since 1965 for the global composite of glaciers. Mean global mass balance over the

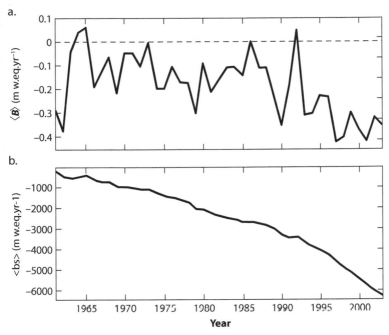

Figure 9.3. Global glacier mass balance, 1961–2004, excluding the Greenland and Antarctic ice sheets. [From Dyurgerov (2002, updated 2005).] The compilation assumes a global glacier area of 785,000 km^2. (a) Mean specific (areally weighted) mass balance rate. (b) Cumulative ice volume change.

full period was −0.24 m yr^{-1} w.e., equivalent to an average rate of global eustatic sea level rise of 0.5 mm yr^{-1}. Glacier contributions to sea level rise have increased markedly since 1990 to an estimated 0.77 mm yr^{-1} from 1991 to 2004 and more than 1 mm yr^{-1} in the first decade of the 2000s.

The Greenland and Antarctic ice sheets are under intense scrutiny. Both are in a state of negative mass balance at the time of writing, with indications that the two ice sheets are more sensitive to climate warming than previously thought. Satellite altimetry and gravity measurements, airborne radar studies of ice thickness, interferometric synthetic aperture radar (INSAR) measurements of ice motion, and microwave mapping of snow facies and the ice sheet ablation area all inform understanding of mass balance and ice-dynamical processes in each ice sheet. The satellite observations provide an unsettling view of recent changes.

Greenland's mass balance has been negative since the early 1990s, after what is believed to be a period of near-balance through much of the 20th century. This is due to a combination of increased surface melting and accelerated ice discharge into the oceans. As a result of these two ablation mechanisms, low altitudes in Greenland have thinned and lost mass since the early 1990s, whereas the interior of the ice sheet, above 2000 m, is near an overall state of balance. Surface mass balance in Greenland is still positive, but regional climate modeling indicates a negative trend in surface mass balance during the past 60 years, with a steepening of melt rates in the 1990s. Recent mass losses due to ice dynamics (calving) and melting in Greenland are similar in magnitude. Greenland is now being carefully monitored to determine whether recent changes reflect a shift to a negative mass balance regime that will persist and perhaps deepen this century or whether the recent extreme years of melting and outlet glacier acceleration may just reflect interannual variability. It is an important and difficult question, as the technology to monitor the ice sheet in detail is relatively new, spanning about two decades, compared with the much longer timescales of many aspects of ice sheet dynamics.

Processes in Antarctica have more in common with those in Greenland than glaciologists have traditionally thought. The Antarctic Peninsula warmed dramatically in the 20th century, eliciting glacial retreat and contributing to the breakup of the Wordie Ice Shelf and the Larsen Ice Shelf. Most of West Antarctica has experienced moderate warming since the 1950s, whereas East Antarctic temperatures have been stable or cooling slightly. Because there is little or no surface melting outside of the Antarctica Peninsula, changes in air temperature are of little consequence to most of the ice sheet. Ice shelves are an important exception to this, as it is believed that surface melting can generate through-going crevasses that weaken ice shelves and contribute to their disintegration. There is still negligible surface melting in the major ice shelves of West Antarctica, the Ross and the Ronne–Filchner, but several degrees Celsius of regional warming could threaten these ice shelves through this mechanism.

Ocean warming probably poses a more immediate threat in Antarctica, as basal melt rates in ice shelves and floating outlet glaciers reach tens of meters per year. Increased dynamical discharge from the Amundsen Sea sector of West Antarctica since the 1990s appears to have been triggered by regional ocean warming, which has caused thinning of marine-based outlet glaciers, inland grounding line migration, and increased iceberg calving and discharge. Where the ice sheet bed has a reverse slope, as is the case in much of West Antarctica, a classical tidewater glacier instability could cause such ice-marginal perturbations to propagate inland for years to decades, accompanied by regional ice sheet thinning and increased flow velocities. Similar observations and processes apply to several major outlet glaciers of the Greenland ice sheet and have driven the recent increase in dynamical ice losses in Greenland.

Increases in high-elevation accumulation are expected to offset partially ice-marginal thinning in both Greenland and Antarctica. Indeed, most climate models forecast that the Antarctic ice sheet will gain mass this century as a result of warming-induced increases in accumulation rate, although it must be emphasized that these models do not include melting at the ice–ocean interface, ice dynamical discharge to the oceans, or potential increases in these sources of ice loss. It is uncertain whether or not snow accumulation has already increased on the Antarctic plateau; available observations are equivocal, but small increases in accumulation may be under way.

Whether increases in accumulation occur in a warmer world depends on the degree to which Antarctic precipitation processes are governed by the Clausius–Clapeyron relation versus cyclonic incursions to the interior plateau. Atmospheric modeling indicates that about half of the snow accumulation in East Antarctica occurs through direct deposition of "diamond dust," with the other half associated with occasional frontal systems that penetrate to the interior of the ice sheet. Accumulation of diamond dust is proportional to specific humidity, but precipitation associated with frontal systems is not directly linked with temperature and humidity. Changes in this source of moisture depend on the patterns and frequency of cyclogenesis and storm tracks in the Southern Ocean.

Changes in Permafrost

Borehole temperature measurements indicate that permafrost and ground temperatures have warmed in the past century. There is also widespread evidence of increasing active layer thickness, melting at the base of permafrost in regions where it is less than ~100 m thick, northward migration of the permafrost zone, and the development of thermokarst terrain and thaw slumps. Because the timescale for surface temperature signals to penetrate to the base of the permafrost is long (ca. 1 year to reach 20 m and 100 years to reach 150 m depth), permafrost thickness and the geographic distribution of permafrost are primarily responding to century- and millennial-scale temperature shifts. Changes in active layer depth, in contrast, can be interpreted as a short-term response to changing temperature and snowpack patterns. Where near-surface permafrost contains massive ice, this surface degradation leads to geomorphologic effects such as thaw slumps and coastal erosion. These and other effects are touched on in the next section.

....

SOCIAL AND ECOLOGICAL EFFECTS OF CRYOSPHERIC CHANGE

Reductions in Arctic sea-ice cover have well-documented effects on polar bear populations and on marine mammals such as ring seals and bowhead whales, which are concentrated near the sea-ice edge. Shifting sea-ice geography and seasonality affects migration, feeding, and breeding patterns for these species. This also impacts circum-Arctic native populations, as the changing ice season affects travel conditions and traditional hunting practices. Changes in the seasonality and extent of open water have unclear implications for marine invertebrates and algal populations. Arctic Ocean ecology is expected to become more akin to that of the adjacent subpolar seas in future decades, although changes will likely be confined to the summer and fall; winter ice cover is still extensive over the Arctic.

Opening of the Arctic for shipping, tourism, and natural resource development is anticipated as summer passage becomes more safe and reliable. This creates commercial possibilities, but it also creates environmental and development challenges for the Arctic region.

Sea, lake, and river ice offer a flat, solid surface that is exploited for ice roads, seasonally connecting remote communities to the "transportation grid." Reduced ice cover has major effects on communities that rely on winter ice roads, as air transportation is the only other way to access these sites for shipping and provision of goods and services. This is an expensive alternative and it leads to isolation. Reductions in river ice cover at lower latitudes also affect the timing and frequency of ice-jam floods, with effects on riparian ecology and flood frequencies. More open water can also lead to increased production of frazil ice, which affects structures and civil engineering works (e.g. frazil buildup over water intakes).

Permafrost degradation has a number of well-documented ecological and societal consequences, in particular effects on infrastructure from ground subsidence and slope failure. Roads, pipelines, and buildings at high latitudes have always faced this challenge, due to the temperature effects of development (i.e. changing land cover), but climate warming is adding to the challenge. The combination of reduced sea ice and sea level rise is giving increased ocean storm swell in some regions of the Arctic, such as the Mackenzie delta, driving high rates of coastal erosion where ground ice is exposed to warm ocean waters.

Active layer deepening and thawing of permafrost are also accompanied by changes in surface hydrology, nutrient cycling, and vegetation cover. Loss of ice alters the water table, and because topography is subtle in many low-lying tundra and peatland areas, surface drainage patterns are evolving rapidly. Soil microbial activity (decomposition) is also accelerated in thawed and warmed ground. Partially compensating for this is an increase in photosynthetic activity and deepening of peat formation in thawed ground, which can lead to stored carbon: an atmospheric sink. Overall, net carbon release to the atmosphere from permafrost thaw is a potentially large positive feedback to future climate change, but there is large uncertainty in this.

Arctic and alpine ecology and hydrology are evolving as a consequence of reduced glacier and permafrost cover. It takes a long time—decades to centuries—for vegetation to move in after the retreat of glaciers, but alpine ecosystems are gradually shifting uphill. A more immediate impact of glacier retreat is the depletion of late-summer streamflow in alpine catchments. Reductions in seasonal snowpack and glacier cover are causing reduced summer streamflow in most alpine environments. This affects downstream communities and in-stream ecology, through both lower flows and increased water temperatures. Water resource management needs to adapt to expected increases in winter and early spring flows, an earlier freshet, and reduced flows from mid-summer to fall in mountain-fed streams.

Sea Level Rise

The loss of ice from mountain glaciers, Greenland, and Antarctica is contributing significantly to sea level rise. Global mean sea levels have risen by approximately 20 cm since 1900, with glacier contributions making up roughly half of this total. Thermal expansion as the oceans warm is the other main driver of sea-level rise. In the first decade of the 21st century, losses in glacier ice and the two continental ice sheets account for about two-thirds of the ca. 33 mm of sea level rise.

Of the world's 197 sovereign nations, 150 (76%) border on an ocean and are therefore directly vulnerable to the gradual swelling of the world's oceans. Sea level rise affects an even larger proportion of global infrastructure and society, as urban centers and populations are concentrated near the world's coastlines. A survey of global cities with populations exceeding 2 million reveals 98 such cities in contact with the ocean, representing a population of about 664 million people at the time of writing; millions more reside in smaller coastal settlements. The process of sea level rise is gradual, so adaptation in developed countries is broadly feasible, but at high costs. Effects are expected to be more severe in low-lying developing countries, where vulnerability to tropical cyclones is high and where populations are likely to be displaced in the coming decades.

Despite recent progress in monitoring and evaluating the rates and sources of sea level rise, future changes in sea level are notoriously difficult to forecast. A number of critical ice-dynamical processes and climate feedbacks are absent or weakly parameterized in current ice sheet models, giving them a conservative bias; they understate the range and extent of potential ice sheet response to climate change. It is therefore difficult to provide an informed estimate of the most likely or most severe sea level rise that can be expected this century. As the century progresses, this may prove to be one of the most critical aspects of cryospheric change.

FUTURE DIRECTIONS

The somber prospect of tens of millions of environmental refugees associated with global sea level rise is not an uplifting note on which to conclude. It does, however, illustrate one profound example of the scale of the cryosphere and its integral role within the global climate system. It also points to an area of intensive current research, as all of the world will be touched by the evolution and fate of the cryosphere in the coming decades and centuries. The atlas of the world will be redrawn as coastal boundaries, navigable waterways, river systems, and population centers shift in the coming decades.

For this reason there are concerted, globally integrated efforts to better understand and constrain cryosphere–climate processes and cryospheric sensitivity to climate change. For the question of sea level rise, this requires improved quantitative understanding of ice sheet dynamics and decadal-scale ice sheet variability. Of particular importance are ice–ocean interactions, fast-flow behavior in ice sheets (i.e., ice streams), and the subglacial processes that enable fast flow.

Fundamental glaciological data are still sparse in large sectors of the polar ice sheets, as well as mountain glaciers, including knowledge of ice thickness, thermal regime, and subglacial conditions. It is difficult to apply hydrological and basal flow models to much of Greenland and Antarctica, where groundwater drainage, sediment properties, and other details of the subglacial environment are poorly known. Methods of parameterizing subgrid-scale basal processes in large-scale models are needed. Similar physical and numerical challenges are involved in simulation of melting beneath floating ice and iceberg calving; there is no established "calving law," as the underlying physics and environmental controls are not fully understood.

Improved coupling between ice sheet and climate models is also necessary. Several research groups have initiated this effort in recent years, but coupled models are still in early stages. Energy- and mass-conserving models that describe ice–ocean coupling will be particularly important going forward and will require regional ocean models that have skill in coastal and sub–ice shelf dynamics.

In general, the mass balance fields simulated by climate models are not accurate enough for coupled ice sheet–climate modeling, nor can sophisticated atmospheric models be integrated over the millennial timescales of interest for ice sheet evolution. However, the development of improved regional-scale meteorological and glacier mass balance models permits direct estimates of surface mass balance from meteorological models and offers a physically based method to simulate how these fields can be expected to change with ice sheet geometry.

The challenge is greater for the mass balance fields and surface climatological forcing of mountain glaciers and snowpacks. Temperature and precipitation gradients are steep in complex terrain. The topography and the relevant meteorological processes are not faithfully resolved in even regional climate models, so some form of climatic downscaling is needed to prescribe mass balance fields for modeling of snow distribution and glacier mass balance in alpine regions. Current downscaling methods generally do not conserve energy or mass, and improved treatments are needed.

There are different considerations for permafrost modeling, but they also relate to the resolution of the landscape and surface climate. The mean annual surface temperature that governs permafrost aggradation and degradation depends on local-scale vegetation, snow-pack depth, hydrology, and soil properties, which commonly vary over a spatial scale of meters. At present, global-scale permafrost models need to be interpreted "statistically" for a region (i.e., via a distribution of permafrost thickness in a given climate-model grid cell, based on the subgrid-scale ground cover and snow conditions in the region).

There is intense interest in understanding the feedbacks associated with sea-ice change, such as effects on polar cloud cover, ocean warming and stratification, and multiyear or decadal-scale sea-ice "memory." Details of sea-ice rheology and the best way to model discontinuous, subgrid-scale processes in continuum models of ocean–ice dynamics are still being explored. Carbon fluxes between sea ice, the ocean mixed layer, and the atmosphere are another subject of great interest; algae and nutrients deposited on sea ice can create impressive plankton blooms when melting delivers these to the ocean, acting as a carbon sink, but the overall effects of changing sea ice on the global carbon budget are not yet clear.

Similarly, deepening active layers in frozen ground and disappearance of permafrost at lower latitudes, along with the geomorphic response to this (e.g., altered hydrological drainage), have an unknown net impact on fluxes of CO_2 and CH_4. Thawing permafrost terrain has the potential to be a significant, large-scale source of carbon to the atmosphere, exacerbating global warming. Insight into this will also help understanding of the potential role of permafrost in the unresolved question of carbon sinks during the glacial period.

Recent efforts have focused on the question of thresholds and reversibility of different aspects of cryospheric response to climate change. This includes aspects of ice sheet and sea-ice retreat, as well as effects of freshwater forcing on ocean circulation. This is an active area of research, focused on nonlinearities in the climate system and concerns about potential long-term commitments in the cryospheric response to anthropogenic forcing of the climate system.

This list of future research directions is far from comprehensive. There are many more priorities within each aspect of cryospheric science. With the notable exception of sea ice, many aspects of the cryosphere have only recently been considered to be fully contributing, interactive components of the global climate system. This means that there are many exciting possibilities to improve coupled cryosphere–climate models at regional and global scales. Satellites are providing a remarkable view of the cryosphere. In many cases, observations are pointing out important new directions for theoretical and field-based studies, and these different strands of research will combine to better illuminate the role of the cryosphere in the global climate system.

Ablation. Snow or ice removal through meltwater runoff, sublimation, wind scour, avalanching, or glacial calving (mechanical fracturing; see *calving*).

Ablation zone. The area of a glacier where annual ablation exceeds annual accumulation, giving net loss of snow and ice.

Accretion. Increase in ice mass by freezing of basal or surface water.

Accumulation. Snow or ice added to an ice mass via snowfall, frost deposition, rainfall that freezes on/in the ice mass, refrozen meltwater, wind-blown snow deposition, or avalanching.

Accumulation area ratio (AAR). The fractional area of a glacier represented by the accumulation zone at the end of the summer melt season (AAR = accumulation zone area/glacier area).

Accumulation zone. The area of a glacier where annual accumulation exceeds annual ablation, giving net accumulation.

Active layer. The layer of seasonally frozen ground in contact with the atmosphere, usually with reference to permafrost terrain. Active layer depths vary from centimeters to several meters.

Aufeis. See *icing*. Also known as *naled*.

Bergeron process. Process by which ice crystals grow at the expense of cloud droplets in *mixed clouds*, through selective vapor deposition onto ice crystals. This process is driven by the lower saturation vapor pressure over ice crystals, relative to water droplets.

Clathrate hydrate. Gases (e.g., methane) trapped in an ice-crystal "shell" or "cage," found frozen in permafrost and in shallow seafloor sediments (e.g., cold continental shelf environments).

Clausius–Clapeyron relation. The thermodynamic relationship between pressure and temperature, which underlies both (i) the pressure melting point depression in ice and (ii) the increase in saturation vapor pressure with temperature in the atmosphere.

Drift ice. See *pack ice*.

East Antarctica. The portion of Antarctica that lies mostly in the Eastern Hemisphere, east of the Transantarctic Mountains (Ross Sea sector) and the Thiel Trough (Weddell Sea sector). Most of the East Antarctic ice sheet is grounded above sea level, on a contiguous landmass.

Equilibrium line altitude (ELA). The elevation at which seasonal snow accumulation balances ablation on a glacier.

Fast ice. Sea ice that is frozen to the shore. Also called landfast ice.

Firn. Multiyear snow that is in transition from meteoric snowfall to glacier ice. Firn densities typically range from 550 to 830 kg m^{-3}.

Floating ice. Ice that is floating in water. Lake, river, and sea ice form from in situ freezing of the water. Once initiated, meteoric precipitation (snow or frozen rainwater) adds to the ice mass. In contrast, floating glacier ice (e.g., in *ice shelves*) is advected/transported to the water body from a terrestrial ice mass. *Icebergs* are fragments of glacier ice that have broken off and are now floating.

Frazil ice. The early stages of small ice crystal growth in rivers, lakes, and oceans, when waters are supercooled and turbulent.

Frost. Deposition of ice on a surface, forming directly from water vapor.

General circulation model (GCM). Three-dimensional model of atmosphere or ocean dynamics. Conveniently, GCM also stands for global climate model, and this acronym is now used interchangeably.

Glaciated. A region or landscape influenced by past glacial (e.g., Pleistocene Ice Age) cover.

Glacier. A perennial terrestrial ice mass that shows evidence of motion via gravitational deformation.

Glacier ice. Polycrystalline ice formed from snow metamorphism, with a density of 830 to 920 kg m^{-3}. The transition from *firn* to ice at ~830 kg m^{-3} is associated with bubble close-off.

Glacierized. A region or river basin where glaciers are currently present.

Glacier mass balance. The overall gain or loss of mass for a glacier or ice sheet over a specified time interval, typically 1 year. This can be expressed as a rate of change of mass (kg yr^{-1}), ice volume (m^3 yr^{-1}), or water-equivalent volume (m^3 w.e. yr^{-1}). See also *specific mass balance* and *surface mass balance*.

Grounded ice. A glacier or ice sheet that is resting on bedrock or sediments. It can be grounded either above or below sea level.

Ground ice. Ice in permafrost or seasonally frozen ground. Also known as soil ice.

Grounding line. The transition zone between grounded and floating ice.

Heinrich event. Episodic surge events from the Hudson Strait sector of the Laurentide ice sheet.

Ice Age. See *Pleistocene Ice Age*.

Iceberg. A fragment of a glacier or ice shelf that has broken off (see *calving*) from the main ice mass and is now floating in a lake or sea.

Ice cap. A dome of glacier ice that overwhelms the local bedrock topography, with the ice flow direction governed by the shape of the ice cap itself.

Icefield. A sheet of glacier ice in an alpine environment in which the ice is not thick enough to overwhelm the local bedrock topography, but is draped over and around it; glacier flow directions in an icefield are dictated by the bed topography.

Ice sheet. A large (i.e., continental-scale) ice cap.

Ice shelf. Glacier ice that has flowed into an ocean or lake and is floating, no longer supported by the bed.

Icing. A thin sheet of ice that forms from refreezing of meltwater (e.g., superimposed ice on glaciers) or from upwelling of groundwater to a cold subaerial environment. Icings can have large horizontal extents of several square kilometers in the latter case. Also known as *aufeis* or *naled*.

Katabatic wind. Downslope wind resulting from gravitational drainage of cold air masses. These are common on valley glaciers and the flanks of ice caps and ice sheets.

Lake ice. Floating ice on a lake, initially formed by freezing of the lake water.

Last glacial maximum (LGM). Period of maximum extent of the last Pleistocene glacial ice sheets, ca. 21,000 years ago.

Massive ice. Lenses or wedges of highly concentrated or pure ice in frozen ground.

Milankovitch cycles. Variations in the Earth–Sun orbit on timescales of 10^4 to 10^5 years (tens to hundreds of kyr), which lead to changing seasonality and latitudinal distribution of insolation. These orbital variations drive the glacial–interglacial cycles of the Quaternary period.

Mixed clouds. Clouds from 0°C to about –40°C, with a mixture of ice crystals, supercooled water droplets, and water vapor.

Névé. The accumulation area of an icefield, often associated with thick layers of firn.

Pack ice. Drifting sea ice that is consolidated.

Pancake ice. Discrete, rounded pieces of sea or lake ice up to a few meters in diameter.

Periglacial. Terrestrial environments influenced by glacial or permafrost activity.

Permafrost. Perennially frozen ground, technically defined as ground that is at or below 0°C for at least 2 years.

Proglacial. The environment adjacent to a glacier, also referred to as the glacier forefield. For most contemporary glaciers, the proglacial environment is the recently deglaciated region where vegetation has yet to take hold.

Orbital variations. See *Milankovitch cycles*.

Pleistocene Ice Age. The last ~2.6 million years in Earth history, characterized by at least 40 advances and retreats of glacial ice (glacial–interglacial cycles) over much of the world, in particular the Northern Hemisphere land mass. Also known as the Quaternary Ice Age.

River ice. Floating ice on a river, initially formed by freezing of the river water.

Sea ice. Floating ice formed by freezing of seawater.

Snow. Ice-crystal precipitation that accumulates on the surface.

Snowball Earth. Episodes of complete global glacier and sea-ice cover in Earth's distant past.

Specific mass balance. The area-averaged mass balance rate on a glacier (kg m^{-2} yr^{-1}), often expressed as the rate of water-equivalent thinning/thickening (m w.e. yr^{-1}).

Subglacial. The environment beneath a glacier, at the ice–bed interface.

Supraglacial. On the surface of a glacier.

Surface mass balance. The mass balance at the atmosphere–glacier interface, associated with net snow accumulation minus surface ablation. This is often referred to as the glacier's mass balance, but strictly speaking mass balance also includes the gain and loss of ice in englacial, subglacial, and ice-marginal environments (i.e., associated with calving).

West Antarctica. The low-lying portion of Antarctica that lies mostly in the Western Hemisphere, west of the Transantarctic Mountains. Most of the West Antarctic ice sheet is grounded below sea level, on a network/ archipelago of islands/tectonic blocks separated by deep troughs.

Notes

CHAPTER 6

1. Based on a global ocean area of $3.62 \times 10^8 \text{ km}^2$ and densities of 910 kg m^{-3} and 1000 kg m^{-3} for ice and meltwater.

2. The time required to reach $(1 - 1/e)$ (~63%) of an exponential transition from an initial to a final state.

CHAPTER 9

1. Irradiance increases with time as part of the ca. 10-billion-year life cycle of G2-type stars like the Sun.

Annotated Bibliography

....

For each chapter, recommendations for further reading are listed followed by literature and Web resources explicitly referenced in the text.

CHAPTER 1: INTRODUCTION TO THE CRYOSPHERE

Further Reading

Gosnell, M. (2005). *Ice: The Nature, the History, and the Uses of an Astonishing Substance.* Alfred A. Knopf, New York, 560 pp.

Slaymaker, O., and R.E.J. Kelly (2007). *The Cryosphere and Global Environmental Change.* Blackwell Publishing, Malden, MA, 261 pp.

United Nations Environment Programme (2007). *Global Outlook for Ice and Snow.* 238 pp. Available at: http://www .unep.org/geo/geo_ice/.

Recent reviews of Antarctica and Greenland provide well-illustrated and detailed assessments of these dominant features in the global cryosphere:

AMAP (2009). *The Greenland Ice Sheet in a Changing Climate: Snow, Water, Ice and Permafrost in the Arctic (SWIPA)* (D. Dahl-Jensen, J. Bamber, C.E. Bøggild, et al.). Arctic

....

Monitoring and Assessment Programme (AMAP), Oslo, Norway, 115 pp.

Turner, J., R.A. Bindschadler, P. Convey, et al. (2009). *Antarctic Climate Change and the Environment*. Scientific Committee on Antarctic Research, Cambridge, UK.

Extensive cryospheric metadata, data, and summary reports are available at the U.S. National Snow and Ice Data Center: http://nsidc.org/index.html.

References Cited

Early energy balance studies identifying the importance of snow and ice albedo to global climate include the seminal works of Budyko and Sellers:

Budyko, M.I. (1969). The effect of solar radiation variations on the climate of the Earth. *Tellus*, 21, 611–619.

Sellers, W.D. (1969). A global climatic model based on the energy balance of the Earth-atmosphere system. *Journal of Applied Meteorology*, 8, 392–400.

Historical sea ice area and extent are from:

Fetterer, F., K. Knowles, W. Meier, and M. Savoie (2002, updated 2011). *Sea Ice Index*. National Snow and Ice Data Center, Boulder, CO (digital media).

Snow cover data from 1966 to 2006 are available from UNEP (2007), compiled following Armstrong and Brodzik (2005). We update this to March 2011 with monthly Northern Hemisphere snow cover data kindly provided by Thomas Estilow of Rutgers University

Global Snow Lab, based on the methods described in Robinson and Frei (2000). See also http://climate.rutgers .edu/snowcover.

Armstrong, R.L., and M.J. Brodzik (2005). Northern Hemisphere EASE-Grid weekly snow cover and sea ice extent version 3. National Snow and Ice Data Center, Boulder, CO (digital media).

Robinson, D.A., and A. Frei (2000). Seasonal variability of northern hemisphere snow extent using visible satellite data. *Professional Geographer*, 51, 307–314.

UNEP/GRID-Arendal (2007). Mean snow-cover extent in the Northern Hemisphere 1966-2006. UNEP/GRID-Arendal Maps and Graphics Library. Available at: http://maps.grida .no/go/ graphic/mean-snow-cover-extent-in-the-northern -hemisphere-1966-2006.

Permafrost area is from Zhang et al. (2008), and glacier and ice sheet extents in table 1.1 are updated from Lemke et al. (2007), with ice on the Antarctic Peninsula accounted as peripheral glacier cover, dynamically distinct from the Antarctic ice sheet. See chapter 6 for a full discussion of ice area and volume estimates in table 1.1.

Brown, J., O.J. Ferrians, Jr., J.A. Heginbottom, and E.S. Melnikov (1998, revised February 2001). *Circum-Arctic Map of Permafrost and Ground-Ice Conditions*. National Snow and Ice Data Center/World Data Center for Glaciology, Boulder, CO (digital media).

Lemke, P., J. Ren, R.B. Alley, I. Allison, J. Carrasco, G. Flato, Y. Fujii, G. Kaser, P. Mote, R.H. Thomas, and T. Zhang (2007). Observations: Changes in snow, ice and frozen ground. In: *Climate Change 2007: The Physical Science Basis*.

Contribution of Working Group I to the Fourth Assessment Report of the Intergovernmental Panel on Climate Change (S. Solomon, D. Qin, M. Manning, Z. Chen, M. Marquis, K.B. Averyt, M. Tignor, and H.L. Miller, eds.). Cambridge University Press, Cambridge, UK, pp. 337–383.

Zhang T., R.G. Barry, K. Knowles, J.A. Heginbottom, and J. Brown (2008). Statistics and characteristics of permafrost and ground ice distribution in the Northern Hemisphere. *Polar Geography*, 31 (1–2), 47–68.

CHAPTER 2: MATERIAL PROPERTIES OF SNOW AND ICE

Further Reading

The microphysics of ice crystals are thoroughly described in Fletcher (1971) and Hobbs (1974). Mellor (1978) and Weeks and Ackley (1982) discuss many aspects of the macroscale behavior of snow and sea ice.

Fletcher, N.H. (1971). Structural aspects of the ice-water system. *Reports on Progress in Physics,* 34, 913–994.

Hobbs, P.V. (1974). *Ice Physics.* Oxford University Press, Oxford, UK, 837 pp.

Libbrecht, K.G. (2005). The physics of snow crystals. *Reports on Progress in Physics,* 68, 855–895.

Mellor, M. (1978). Engineering properties of snow. *Journal of Glaciology*, 19 (81), 15–66.

Weeks, W.F., and S.F. Ackley (1982). *The Growth, Structure, and Properties of Sea Ice.* CRREL Monograph 82-1. U.S. Army Corps of Engineers, 130 pp.

References Cited

Gosnell's (2005) popular treatise on *Ice* (cf. chapter 1) offers a well-crafted and accessible discussion of some of the peculiarities of water, including the delightful quote from James Trefil.

Trefil, J. (1986). *Meditations at 10,000 Feet: A Scientist in the Mountains*. Charles Scribner's Sons, New York.

Those seeking a more physics-based discussion of the crystal structure of snowflakes will find reward in Fletcher (1971) and Libbrecht (2005). Libbrecht provides some stunning photographs of synthetic crystals. There is also an extensive and impressive collection of work on scanning electron microscope images of snowflakes based out of the U.S. Department of Agriculture labs in Beltsville, Maryland (e.g., Wergin et al., 1998; Rango et al., 2000). Bill Wergin kindly provided the microphotograph in figure 2.1b.

Rango, A., W.P. Wergin, E.F. Erbe, and E.G. Josberger (2000). Snow crystal imaging using scanning electron microscopy. III. Glacier ice, snow and biota. *Journal of Hydrological Sciences*, 45 (3), 357–375.

Wergin, W.P., A. Rango, and E.F. Erbe (1998). Image comparisons of snow and ice crystals photographed by light (video) microscopy and low temperature scanning electron microscopy. *Scanning*, 20, 285–296.

Different values for the thermodynamic properties of snow and ice are found in the research literature, so it is difficult to distill this to single "recommended" values.

Consistent with the climate-system emphasis within this text, I emphasize the macroscale (rather than molecular-scale) thermodynamic behavior, and recommended values here are based on recent field-data compilations, where possible.

Snow density and snow/ice albedo data from Haig Glacier in the Canadian Rockies (figures 2.2 and 2.4) are from unpublished data collected by the author. The field site, instrumentation, and snow sampling methods are described in:

Shea, J.M., F.S. Anslow, and S.J. Marshall (2005). Hydrome-teorological relationships on the Haig Glacier, Alberta, Canada. *Annals of Glaciology*, 40, 52–60.

Sinclair, K.E. and S.J. Marshall (2009). The impact of vapour trajectory on the isotope signal of Canadian Rocky Mountain snowpacks. *Journal of Glaciology,* 55 (191), 485–498.

Values for the density of firn and ice are from Cuffey and Paterson (2010). Herron and Wu (1994) discuss the effects of solar radiation on subsurface melt and the resulting effects on lake ice density during late stages of melt.

Cuffey, K.M., and W.S.B. Paterson (2010). *The Physics of Glaciers*, 4th ed. Butterworth-Heinemann, Oxford, UK, 693 pp.

Herron, R., and M.K. Woo (1994). Decay of a High Arctic lake-ice cover: Observations and modelling. *Journal of Glaciology*, 40 (135), 283–292.

The effects of dissolved air bubbles, impurities, and pressure on the melting point of glacier ice are discussed in detail in Cuffey and Paterson (2010). Parameterizations for the thermal conductivity and heat capacity of glacier ice are also adopted from Cuffey and Paterson (2010). For sea ice and seasonal snow, these thermodynamic values are synthesized from:

Bitz, C.M., and W.H. Lipscomb (1999). An energy-conserving thermodynamic model of sea ice. *Journal of Geophysical Research*, 104, 15669–15677.

Pringle, D.J., H. Eicken, H.J. Trodahl, and L.G.E. Backstrom (2007). Thermal conductivity of landfast Antarctic and Arctic sea ice. *Journal of Geophysical Research*, 112, C04017, doi:10.1029/2006JC003641.

Sturm, M., J. Holmgren, M. König, and K. Morris (1997). The thermal conductivity of seasonal snow. *Journal of Glaciology*, 43 (143), 26–41.

Sturm, M., D.K. Perovich, and J. Holmgren (2002). Thermal conductivity and heat transfer through the snow on the ice of the Beaufort Sea. *Journal of Geophysical Research*, 107 (C10), 8047, doi:10.1029/2000JC000400.

Untersteiner, N. (1961). On the mass and heat budget of Arctic sea ice. *Arch. Meteorol. Geophys. Bioklimatol. Ser. A.*, 12, 151–182.

Untersteiner, N. (1964). Calculations of temperature regime and heat budget of sea ice in the Central Arctic. *Journal of Geophysical Research*, 69, 4755–4766.

Anne Nolin kindly provided the data for spectral reflectance of snow as a function of wavelength, based on the model of Wiscombe and Warren (1980). Typical snow and ice albedo values and the physical properties that affect these are gathered from:

Allison, I., R.E. Brandt, and S.G. Warren (1993). East Antarctic sea ice: Albedo, thickness distribution, and snow cover. *Journal of Geophysical Research*, 98 (C7), 12417–12429.

Grenfell, T.C., S.G. Warren, and P.C. Muller (1994). Reflection of solar radiation by the Antarctic snow surface at ultraviolet, visible, and near-infrared wavelengths. *Journal of Geophysical Research*, 99 (D9), 18699–18684.

Wiscombe, W.J., and S.G. Warren (1980). A model for the spectral albedo of snow, I: Pure snow. *Journal of Atmospheric Science*, 37, 2712–2733.

Warren, S.G., and W.J. Wiscombe (1980). A model for the spectral albedo of snow, II: Snow containing atmospheric aerosols. *Journal of Atmospheric Science*, 37, 2734–2745.

CHAPTER 3: SNOW AND ICE THERMODYNAMICS

Further Reading

Numerous texts describe the essentials of atmospheric thermodynamics, which provide helpful complementary reading. Two excellent examples include:

Bohren, C.F., and B.A. Albrecht (1998). *Atmospheric Thermodynamics*. Oxford University Press, New York.

Petty, G. (2008). *Introduction to Atmospheric Thermodynamics*. Sundog Press, Madison, WI.

The text of Stull (1988) offers a thorough general treatment of boundary layer meteorology, which lays out the theoretical foundation for turbulent energy exchange.

Specific applications to turbulent heat transfer over snow and ice are presented in Andreas (1987, 2002).

Andreas, E.L. (1987). A theory for the scalar roughness and the scalar transfer coefficients over snow and sea ice. *Boundary Layer Meteorology*, 38, 159–184.

Andreas, E.L. (2002). Parameterizing scalar transfer over snow and ice: A review. *Journal of Hydrometeorology*, 3, 417–432.

Stull, R.B. (1988). *An Introduction to Boundary Layer Meteorology*. Kluwer Academic Publishers, Dordrecht, The Netherlands, 666 pp.

References Cited

Calculations in this chapter draw from the thermal properties of snow and ice discussed in chapter 2, with the discussion of sea-ice thermodynamics based on Bitz and Lipscomb (1999). This has its roots in the classical model of Maykut and Untersteiner (1971). Bitz and Marshall (2011) extend this to a more general treatment of cryospheric thermodynamics.

Bitz, C.M., and S.J. Marshall (2011). Cryosphere models—ocean and land. In: *Encyclopedia of Sustainability Science and Technology* (P. Rasch, ed.). Springer, New York.

Maykut, G.A., and N. Untersteiner (1971). Some results from a time-dependent thermodynamic model of sea ice. *Journal of Geophysical Research*, 76, 1550–1575.

The equations for modeling of shortwave radiation are outlined in Garnier and Ohmura (1968) and Oke (1987).

Garnier, B.J., and A. Ohmura (1968). A method of calculating the direct short-wave radiation income of slopes. *Journal of Applied Meteorology*, 7 (5), 796–800.

Oke, T.R. (1987). *Boundary Layer Climates*, 2nd ed. Methuen, London, 435 pp.

There is a plethora of papers discussing energy balance calculations over snow and ice. Some excellent examples include:

Arnold, N.S., I.C. Willis, M.J. Sharp, K.S. Richards, and M.J. Lawson (1996). A distributed surface energy-balance model for a small valley glacier. I. Development and testing for Haut Glacier d'Arolla, Valais, Switzerland. *Journal of Glaciology*, 42 (140), 77–89.

Cline, D.W. (1997). Snow surface energy exchanges and snowmelt at a continental, midlatitude Alpine site. *Water Resources Research*, 33 (4), 689–701.

Marks, D., J. Domingo, D. Susong, T. Link, and D. Garen (1999). A spatially distributed energy balance snowmelt model for application in mountain basins. *Hydrological Processes*, 16, 1935–1959.

Surface energy balance field data from Kwadacha Glacier was collected by the author (unpublished). The field site and automatic weather station data from this site are discussed in:

Bolch, T., B. Menounos, and R. Wheate (2010). Landsat-based inventory of glaciers in western Canada, 1985-2005. *Remote Sensing of the Environment*, 114, 127–137.

Losic, M. (2009). On the turbulent heat flux contributions to energy balance at Opabin Glacier, Yoho National Park, Canada. Unpublished M.Sc. thesis, University of Calgary.

CHAPTER 4: SEASONAL SNOW AND FRESHWATER ICE

Further Reading

Mariana Gosnell's *Ice* (see chapter 1) provides an engaging narrative on the subtleties and quirks of lake ice. References cited below provide more scientific detail on freshwater ice. Gray and Male's (1981) *Handbook of Snow* gives an expansive discussion into many aspects of seasonal snow that are not possible to discuss here, with a focus on snow hydrology. Kenneth Libbrecht offers insights and a remarkable collection of photographs capturing the wonder and intricacies of snow crystals, with a repertoire of texts that range from children's books to advanced physics.

Gray, D.M., and D.H. Male, eds. (1981). *Handbook of Snow: Principles, Processes, Management and Use.* Pergamon Press, Toronto, Canada, 776 pp.

Libbrecht, K.G. (2003). *The Snowflake: Winter's Secret Beauty.* Voyageur Press, Stillwater, ME.

Libbrecht, K.G. (2005). The physics of snow crystals. *Reports on Progress in Physics,* 68 (4), 855–895.

References Cited

Quotations from Emerson and Helprin capture some aspects of the magic and character of snow. The sampling of Inuit terms for snow and ice in table 4.1 also conveys this. These are cobbled from various sources, including

Cone (2005). I base my interpretation of *sikussaq* on the descriptions of Koch (1926).

Cone, M. (2005). *Silent Snow: The Slow Poisoning of the Arctic.* Grove/Atlantic Press, New York, 256 pp.

Emerson, R.W. (1898). *Poems. New and Revised Edition.* The Riverside Press, Cambridge, MA.

Helprin, M. (1983). *Winter's Tale.* Harcourt Books, Orlando, FL, 672 pp.

Koch, L. (1926). Ice cap and sea ice in North Greenland. *Geographical Review,* 16 (1), 98–107.

Snow cover statistics are from the Rutgers University Global Snow Lab (see chapter 1). The physics of snow and ice melt are discussed in chapter 3, with simplifications to the energy balance that are commonly used to model snow melt discussed in Hock (1999) and Ohmura (2001). Herron and Woo (1994) examine some features specific to lake ice decay (see chapter 3).

Hock, R. (1999). A distributed temperature-index ice- and snowmelt model including potential direct solar radiation. *Journal of Glaciology,* 45 (149), 101–111.

Ohmura, A. (2001). Physical basis for the temperature/melt-index method. *Journal of Applied Meteorology,* 40, 753–761.

Saturation vapor expressions for snow and ice are from:

World Meteorological Organization (2008). *Guide to Meteorological Instruments and Methods of Observation,* Appendix 4B. WMO-No. 8 (CIMO Guide). World Meteorological Organization, Geneva, Switzerland.

This chapter draws from the following papers on river and lake ice:

Adams, W.P. (1981). Snow and ice on lakes. In: *Handbook of Snow* (D.M. Gray and D.H. Male, eds.). Pergamon Press, Toronto, Canada, pp. 437–474.

Ashton, G.D. (1978). River ice. *Annual Review of Fluid Mechanics*, 10, 369–392.

Duguay, C.R., G.M. Flato, M.O. Jeffries, P. Ménard, K. Morris, and W.R. Rouse (2003). Ice-cover variability on shallow lakes at high latitudes: Model simulations and observations. *Hydrological Processes*, 17, 3465–3483.

Martin, S. (1981). Frazil ice in rivers and oceans. *Annual Review of Fluid Mechanics*, 13, 379–397.

Prowse, T.P. (1995). River ice processes. In: *River Ice Jams* (S. Beltaos, ed.). Water Resources Publications, LLC, Highlands Ranch, CO, pp. 29–70.

Meteorological data for Yellowknife and Eureka, northern Canada, are from Environment Canada (http://www.climate.weatheroffice.gc.ca/climateData/canada_e.html). Estimates of the load capacity of lake ice are synthesized from various (sometimes conflicting) sources, including *A Field Guide to Ice Construction Safety* (2007), published by the Northwest Territories Department of Transportation (http://www.dot.gov.nt.ca), the State of Minnesota Department of natural Resources (http://www.dnr.state.mn.us/safety/ice/thickness.html), and the Finnish Road Administration.

CHAPTER 5: SEA ICE

Further Reading

The classical text edited by Norbert Untersteiner, *The Geophysics of Sea Ice*, contains many fundamentals that shape current understanding and modeling of sea ice, including several topics that were not possible to cover here. Several other seminal articles are recommended here.

Ackley, S.F., and W.F. Weeks (1990). *Sea Ice Properties and Processes*. CRREL Monograph 90-1. U.S. Army Corps of Engineers.

Hibler, W.D., III (1979). A dynamic thermodynamic sea ice model. *Journal of Physical Oceanography*, 9, 815–846.

Maykut, G.A., and N. Untersteiner (1971). Some results from a time-dependent thermodynamic model of sea ice. *Journal of Geophysical Research*, 76, 1550–1575.

Hunke, E.C., and J.K. Dukowicz (1997). An elastic-viscous-plastic model for sea ice dynamics. *Journal of Physical Oceanography*, 27, 1849–1867.

Thorndike, A.S., D.S. Rothrock, G.A. Maykut, and R. Colony (1975). The thickness distribution of sea ice. *Journal of Geophysical Research*, 80, 4501–4513.

Untersteiner, N., ed. (1986). *The Geophysics of Sea Ice*. NATO ASI Series B, Vol. 146. Plenum Press, New York, 1196 pp.

Numerous texts examine ocean physics and the properties of seawater: essential background reading for those who wish to delve further into sea-ice dynamics. Pond and Pickard (1991) is a very readable introduction. The text by Vallis (2012) in this series also offers a fresh and accessible overview.

Pond, S., and G.L. Pickard (1991). *Introductory Dynamical Oceanography*, 2nd ed. Pergamon Press, Oxford, UK, 329 pp.

Vallis, G. (2012). *Climate and the Oceans*. Princeton Primers in Climate. Princeton University Press, Princeton, NJ.

References Cited

Quotations in the chapter are from:

Cuncliffe, B.W. (2002). *The Extraordinary Voyages of Pytheas the Greek*. Penguin Books, New York, 183 pp.

Nansen, F. (1897). *The First Crossing of Greenland*. Reprinted 2001, University Press of the Pacific, Honolulu, HI, 452 pp. Quote from p. 83.

The freezing point of seawater as a function of salinity is given in most of the references above (e.g., Untersteiner, 1986). I adopt the treatment of sea-ice thermodynamics from Maykut (1978), Bitz and Lipscomb (1999), and Bitz and Marshall (2011) (see chapter 2). Hemispheric sea-ice extent and area are from Fetterer et al. (2002, updated 2010) (see chapter 1). Arctic ice-thickness data in figure 5.2a is based on Icesat satellite altimeter analyses of Kwok et al. (2009), available at http://rkwok.jpl.nasa.gov/icesat/download.html. Antarctic ice thickness fields in figure 5.2b are model-generated, adapted from Timmermann et al. (2009). Timmermann et al. (2004) discuss the spatial patterns of Antarctic ice thickness based on the available long-term data.

Kwok, R., G.F. Cunningham, M. Wensnahan, I. Rigor, H.J. Zwally, and D. Yi (2009). Thinning and volume loss of the Arctic Ocean sea ice cover: 2003-2008. *Journal of Geophysical Research*, 114 (C070005), doi:10.1029/2009JC005312.

Maykut, G.A. (1978). Energy exchange over young sea ice in the central arctic. *Journal of Geophysical Research,* 83 (C7), 3646–3658.

Timmermann, R., A. Worby, H. Goosse, and T. Fichefet (2004). Utilizing the ASPeCt sea ice thickness dataset to evaluate a global coupled sea ice-ocean model. *Journal of Geophysical Research*, 109 (C07017), doi:10.1029/2003JC002242.

Timmermann, R., S. Danilov, J. Schröter, C. Böning, D. Sidorenko, and K. Rollenhagen (2009). Ocean circulation and sea ice distribution in a finite element global sea ice-ocean model. *Ocean Modelling*, 27 (3–4), 114–129.

CHAPTER 6: GLACIERS AND ICE SHEETS

Further Reading

The classical text of W.S.B. Paterson, *The Physics of Glaciers*, remains the best treatment of glacier physics, striking a nice balance between theory, process, and empirical insight. Cuffey and Paterson (2010) offer an update on this text (see chapter 2). Hooke (2005) and van der Veen (1999) provide alternative treatments, focused on glacier mechanics. The recent contribution in this genre from Greve and Blatter (2009) is an excellent presentation of the mathematical and continuum mechanical framework for glacier dynamics. Clarke (2005) is recommended for an overview of subglacial processes, one of the most poorly understood aspects of cryospheric science. Oerlemans (2001) gives a nice overview of glacier-climate processes.

Clarke, G.K.C. (2005). Subglacial processes. *Annual Review of Earth and Planetary Sciences*, 33, 247–276.

Greve, R., and H. Blatter (2009). *Dynamics of Ice Sheets and Glaciers*. Springer-Verlag, Berlin, Germany.

Hooke, R. LeB. (2005). *Principles of Glacier Mechanics*, 2nd ed. Cambridge University Press, Cambridge, UK.

Oerlemans, J. (2001). *Glaciers and Climate Change*. Balkema, Rotterdam, The Netherlands.

Paterson, W.S.B. (1994). *The Physics of Glaciers*, 3rd ed. Elsevier, Amsterdam, The Netherlands.

van der Veen, C.J. (1999). *Fundamentals of Glacier Dynamics*. AA Balkema, Rotterdam, The Netherlands.

References Cited

Glacier number, area, and volume estimates come from a variety of sources. Detailed regional information is available in the United States Geological Survey (USGS) satellite atlas of Williams and Ferrigno (1988), with 11 volumes covering different parts of the world. These volumes have been released gradually across the period 1988–2010, and the series is now nearly complete. Dyurgerov and Meier (2005) estimate the global glacier count to be 160,000, of which about 123,000 have some amount of archived information in the World Glacier Inventory. The number of glaciers is understood to be considerably larger than this, however, as thousands of small ice masses are uncounted; estimates from 200,000 to 400,000 have been suggested. It can be assumed that there at least 200,000 glaciers worldwide.

The area of ice in Alaska–Yukon is from Berthier et al. (2010). Arctic ice cap areas are from Dowdeswell and Hagen (2004) and Dyurgerov and Carter (2004). Glaciers and icefields adjacent to the Greenland and Antarctic ice sheets are based on the estimates of Weidick and Morris (1998) and Dyurgerov and Meier (2005). These are a matter of some controversy, as the ice masses are not well mapped and are often indistinct from the ice sheets. The USGS Satellite Atlas for Greenland (Volume C, 1995) gives an estimate of 48,600 km^2 for the glaciers peripheral to the Greenland ice sheet, but a value of 70,000 km^2 is commonly adopted (Weidick and Morris, 1998; Dyurgerov and Meier, 2005). Radić and Hock (2010) adopt 54,000 km^2 for Greenland's peripheral ice masses. I give a "median" estimate of 60,000 km^2 here. Reported values for Antarctica range from 70,000 km^2 (Weidick and Morris, 1998) to 175,000 km^2 (Dyurgerov and Meier, 2005). The latter is adapted from Shumskiy (1969), who estimated 169,000 km^2. Radić and Hock (2010) also use Shumskiy's estimate. I use this value but add in 3700 km^2 from the sub-Antarctic islands along with 300,400 km^2 from icefields of the Antarctic Peninsula, which are geographically and dynamically distinct from the West Antarctic and East Antarctic ice sheets. This estimate for the

Antarctic Peninsula is from the USGS Satellite Atlas for Antarctica (Volume B, 1988, after Drewry et al., 1982).

Berthier, E., E. Schiefer, G.K.C. Clarke, B. Menounos, and F. Rémy (2010). Contribution of Alaskan glaciers to sea-level rise derived from satellite imagery. *Nature Geoscience*, 3 (2), 92–95.

Dowdesell, J.A., and J.O. Hagen (2004). Arctic glaciers and ice caps. In: *Mass Balance of the Cryosphere* (J.L. Bamber and A.J. Payne, eds.). Cambridge University Press, Cambridge, UK, pp. 527–557.

Drewry, D.J., S.R. Jordan, and E. Jankowski (1982). Measured properties of the Antarctic ice sheet: Surface configuration, ice thickness, volume, and bedrock characteristics. *Annals of Glaciology*, 3, 83–91.

Dyurgerov, M.B., and C.L. Carter (2004). Observational evidence of increases in freshwater inflow to the Arctic Ocean. *Arctic, Antarctic, and Alpine Research*, 36 (1), 117–122.

Dyurgerov, M.B., and M.F. Meier (2005). *Glaciers and the Changing Earth System: A 2004 Snapshot*. INSTAAR Occasional Paper 58. Institute of Arctic and Alpine Research, University of Colorado, Boulder, CO, 118 pp.

Radić, V., and R. Hock (2010). Regional and global volumes of glaciers derived from statistical upscaling of glacier inventory data. *Journal of Geophysical Research*, 115 (F01010), doi:10.1029/2009JF001373.

Shumskiy, P.A. (1969). Glaciation. In: *Atlas of Antarctica*, Vol. 2 (E. Tolstikov, ed.). Hydrometeoizdat, Leningrad, pp. 367–400.

Weidick, A., and E. Morris (1998). Local glaciers surrounding continental ice sheets. In: *Into the Second Century of*

World Glacier Monitoring—Prospects and Strategies. A contribution to the IHP and the GEMS (W. Haeberli, M. Hoelzle, and S. Suter, eds.). Prepared by the World Glacier Monitoring Service. UNESCO Publishing, Paris, France, pp. 197–207.

Williams, R.S., Jr., and J.G. Ferrigno, eds. (1988). *Satellite Image Atlas of Glaciers of the World*. U.S. Geological Survey Professional Paper 1386, Volumes A–K. U.S. Geological Survey.

Estimates of ice volume and sea level equivalence for the Greenland and Antarctic ice sheets are from Bamber et al. (2001) and Lythe et al. (2001), respectively. West Antarctic values are updated based on Bamber et al. (2009). An update of Antarctic ice thickness and bedrock topography will soon be available from the BEDMAP2 project. Estimates of the fraction of each ice sheet grounded below sea level are also from these data sets.

Bamber, J.L., R.L. Layberry, and S.P. Gogenini (2001). A new ice thickness and bed data set for the Greenland ice sheet 1: Measurement, data reduction, and errors. *Journal of Geophysical Research*, 106 (D24), 33773–33780. (Data provided by the National Snow and Ice Data Center DAAC, University of Colorado, Boulder, CO. Available at: http://nsidc .org/data/nsidc-0092.html.)

Bamber, J.L., R.E.M. Riva, B.L.A. Vermeersen, and A.M. LeBrocq (2009). Reassessment of the potential sea-level rise from a collapse of the West Antarctic Ice Sheet. *Science*, 324 (5929), 901–903.

Lythe, M.B., D.G. Vaughan, and the BEDMAP Consortium (2001). A new ice thickness and subglacial topographic model of Antarctica. *Journal of Geophysical Research*, 106 (B6), 11335–11351. (Available from the Goddard Space Flight Center Global Change Master Directory: http://gcmd.nasa .gov/index.html.)

Volume estimates for mountain glaciers and polar ice caps are from Dyurgerov and Meier (2005), Radić and Hock (2010), and the sources listed below. Values in table 6.1 are averaged from this compilation of sources, with icefields of the Antarctic Peninsula added in. I adopt an ice volume of 0.095 km^3 (0.24 msl) for the Antarctic Peninsula, from Bamber et al. (2009).

Dyurgerov, M.B. (2002, updated 2005). *Glacier Mass Balance and Regime: Data of Measurements and Analysis* (M. Meier and R. Armstrong, eds.). INSTAAR Occasional Paper 55. Institute of Arctic and Alpine Research, University of Colorado, Boulder, CO.

Ohmura, A. (2004). Cryosphere during the twentieth century. In: *The State of the Planet: Frontiers and Challenges in Geophysics* (R.S.J. Sparks and C.J. Hawkesworth, eds.). Geophysical Monograph 150. International Union of Geodesy and Geophysics, Boulder, CO, and American Geophysical Union, Washington, DC, pp. 239–257.

Raper, S.C.B., and R.J. Braithwaite (2005). The potential for sea level rise: New estimates from glacier and ice cap area and volume distribution. *Geophysical Research Letters*, 32 (L05502), doi:10.1029/2004GL021981.

Recent area changes for glaciers in the Alps and in western Canada are from Paul et al. (2004) and Bolch et al. (2010) (see chapter 3), respectively.

....

Paul, F., A. Kääb, M. Maisch, T. Kellenberger, and W. Haeberli (2004). Rapid disintegration of Alpine glaciers observed with satellite data. *Geophysical Research Letters*, 31 (L21402), doi:10.1029/2004GL020816.

Papers discussing ENSO influences on glacier mass balance in western North America include Bitz and Battisti (1999) and Hodge et al. (1999). Nesje et al. (2000) examine the influence of the North Atlantic Oscillation on Scandinavian glaciers. Oerlemans (1992, 1997) discusses the climate sensitivity of glaciers in Norway and New Zealand. Marshall et al. (2011) consider glaciers in the Canadian Rockies.

Bitz, C.M., and D.S. Battisti (1999). Interannual to decadal variability in climate and the glacier mass balance in Washington, Western Canada, and Alaska. *Journal of Climate*, 12, 3181–3196.

Hodge, S.M., D.C. Trabant, R.M. Krimmel, T.A. Heinrichs, R.S. March, and E.G. Josberger. (1999). Climate variations and changes in mass of three glaciers in western North America. *Journal of Climate*, 11 (9), 2161–2179.

Marshall, S.J., E.C. White, M.N. Demuth, T. Bolch, R. Wheate, B. Menounos, M. Beedle, and J.M. Shea (2011). Glacier water resources on the eastern slopes of the Canadian Rocky Mountains. *Canadian Water Resources Journal*, 36 (2), 109–134.

Nesje, A., Ø. Lie, and S.O. Dahl (2000). Is the North Atlantic Oscillation reflected in Scandinavian glacier mass balance records? *Journal of Quaternary Science*, 15 (6), 587–601.

Oerlemans, J. (1992). Climate sensitivity of glaciers in southern Norway: Application of an energy-balance model to

Nigardsbreen, Hellstugubreen and Alfolbreen. *Journal of Glaciology*, 38 (129), 223–232.

Oerlemans, J. (1997). Climate sensitivity of Franz Joseph Glacier, New Zealand, as revealed by numerical modelling. *Arctic and Alpine Research*, 29, 233–239.

The discussion of ice rheology is based on:

Duval, P. (1981). Creep and fabrics of polycrystalline ice under shear and compression. *Journal of Glaciology*, 27, 129–140.

Glen, J.W. (1955). The creep of polycrystalline ice. *Proceedings of the Royal Society of London, Series A,* 228, 519–538.

Nye, J.F. (1953). The flow law of ice from measurements in glacier tunnels, laboratory experiments, and the Jungfraufirn borehole experiment. *Proceedings of the Royal Society of London, Series A,* 219, 477–489.

Nye, J.F. (1957). The distribution of stress and velocity in glaciers and ice sheets. *Proceedings of the Royal Society of London, Series A,* 275, 87–112.

CHAPTER 7: PERMAFROST

Further Reading

I have neglected many interesting geomorphic processes in frozen ground because they have little role to play in the climate system. This includes intriguing features like pingos, ice wedges, solifluction lobes, and patterned ground. Interested readers can find a wealth of information in the texts of Williams and Smith (1989) and French and Williams (2007).

French, H.M., and P.J. Williams (2007). *The Periglacial Environment*, 3rd ed. Wiley, Chichester, UK.

Williams, P.J., and M.W. Smith (1989). *The Frozen Earth: Fundamentals of Geocryology.* Cambridge University Press, Cambridge, UK.

References Cited

Values for thermal diffusivity are adapted from Williams and Smith (1989) and Slaymaker and Kelly (2007) (see chapter 1). Northern Hemisphere permafrost area is from Zhang et al. (2008) and the map data for the figure are from Brown et al. (1998, updated 2001), also given in chapter 1. Mackay (1972) and Osterkamp (2001) discuss subsea permafrost. Modeling of permafrost thermodynamics is discussed by Osterkamp (1987) and Romanovsky et al. (1997). Zhang (2005) reviews the thermal effects of snow cover on permafrost and active layer thickness. Borehole temperature records in Alaska are quoted from Lachenbruch and Marshall (1986). Lemke et al. (2007) (see chapter 1) discuss the larger-scale record of recent climate change documented in borehole temperatures in permafrost.

Lachenbruch, A.H., and B.V. Marshall (1986). Changing climate: Geothermal evidence from permafrost in the Alaskan Arctic. *Science*, 234, 689–696.

Mackay, J.R. (1972). Offshore permafrost and ground ice, southern Beaufort Sea, Canada. *Canadian Journal of Earth Science*, 9, 1550–1561.

Osterkamp, T.E. (1987). Freezing and thawing of soils and permafrost containing unfrozen water or brine. *Water Resources Research*, 23 (12), 2279–2285.

Osterkamp, T.E. (2001). Sub-sea permafrost. In: *Encyclopedia of Ocean Sciences*. Academic Press, San Diego, CA, pp. 2902–2912.

Romanovsky, V.E., T.E. Osterkamp, and N.S. Duxbury (1997). An evaluation of three numerical models used in simulations of the active layer and permafrost temperature regimes. *Cold Regions Science and Technology*, 26 (3), 195–203.

Zhang, T. (2005). Influence of the seasonal snow cover on the ground thermal regime: An overview. *Reviews of Geophysics*, 43, RG4002, doi:10.1029/2004RG000157.

CHAPTER 8: CRYOSPHERE–CLIMATE PROCESSES

Further Reading

The role of the cryosphere in the global climate system is thoroughly discussed in the Intergovernmental Panel on Climate Change (IPCC) reports [e.g., Lemke et al. (2007), cited in chapter 1]. A dedicated discussion is also offered in Barry (2002). Serreze and Barry (2005) cover many aspects of surface energy balance over snow and ice and large-scale (Arctic) cryosphere–climate processes.

Barry, R.G. (2002). The role of snow and ice in the global climate system: A review. *Polar Geography*, 24 (3), 235–246.

Serreze, M.C., and R.G. Barry (2005). *The Arctic Climate System*. Cambridge University Press, Cambridge, UK.

References Cited

The 1D global energy balance model is adapted from the models of Budyko (1969) and Sellers (1969), as cited in chapter 1. Values cited for the global energy budget are based on Trenberth (2009) and Levitus et al. (2009). Statistics for global energy consumption are from http://www.bp.com/statisticalreview. Chapter 2 of the 2007 IPCC scientific assessment, by Forster et al. (2007), gives estimates of current anthropogenic radiative forcing. Romanova et al. (2006) present GCM studies of the effects of snow and ice cover on the global albedo.

Forster, P., V. Ramaswamy, P. Artaxo, et al. (2007). Changes in atmospheric constituents and in radiative forcing. In: *Climate Change 2007: The Physical Science Basis. Contribution of Working Group I to the Fourth Assessment Report of the Intergovernmental Panel on Climate Change* (S. Solomon, D. Qin, M. Manning, Z. Chen, M. Marquis, K.B. Averyt, M. Tignor, and H.L. Miller, eds.). Cambridge University Press, Cambridge, UK, pp. 129–234.

Levitus, S., J.I. Antonov, T.P. Boyer, R.A. Locarnini, H.E. Garcia, and A.V. Mishonov. (2009). Global ocean heat content 1955-2008 in light of recently revealed instrumentation problems. *Geophysical Research Letters*, 36 (L07608), doi:10.1029/2008GL037155.

Romanova, V., G. Lohmann, and K. Grosfeld (2006). Effect of land albedo, CO_2, orography, and oceanic heat transport on extreme climates. *Climate of the Past*, 2, 31–42.

Trenberth, K.E. (2009). An imperative for adapting to climate change: Tracking Earth's global energy. *Current Opinion in Environmental Sustainability*, 1, 19–27.

The discussion of ice–ocean interactions draws from several sources. Great salinity anomalies in the North Atlantic are discussed by Dickson et al. (1988) and Belkin et al. (1998). Hemming (2004) reviews North Atlantic Heinrich events from the glacial period. The impact of ocean variability on ice sheet dynamics is a subject of much attention at the time of writing, motivated by observations of exceptional sensitivity of outlet glaciers to ocean conditions (e.g., Rignot and Jacobs, 2002; Rignot and Kanagaratnam, 2006; Holland et al., 2008). I discuss this further in chapter 9.

Belkin, I.M., S. Levitus, J. Antonov, and S.-A. Malmberg (1998). "Great Salinity Anomalies" in the North Atlantic. *Progress in Oceanography*, 41 (1), 1–68.

Dickson, R.R., J. Meincke, S.-A. Malmberg, and A.J. Lee (1988). The "Great Salinity Anomaly" in the northern North Atlantic, 1968-1982. *Progress in Oceanography*, 20, 103–151.

Hemming, S.R. (2004). Heinrich events: Massive late Pleistocene detritus layers of the North Atlantic and their global imprint. *Reviews of Geophysics*, 42 (RG1005), doi:10.1029/2003RG000128.

Holland, D.M., R.H. Thomas, B. De Young, M.H. Ribergaard, and B. Lyberth (2008). Acceleration of Jakobshavn Isbrae triggered by warm subsurface ocean waters. *Nature Geoscience*, 1, 659–664.

Rignot, E., and S. Jacobs (2002). Rapid bottom melting widespread near Antarctic Ice Sheet grounding lines. *Science*, 296 (5575), 2020–2023.

Rignot, E., and P. Kanagaratnam (2006). Changes in the velocity structure of the Greenland ice sheet. *Science*, 311, 986–990.

Schuur et al. (2008) give an overview of carbon sources and sinks in permafrost and the likely response of this carbon reservoir to climate warming. Turetsky et al. (2002, 2007) examine carbon fluxes from boreal peatlands, where most of the permafrost carbon is stored. Walter et al. (2006) discuss methane release from thawing of wetlands.

Schuur, E.A.G., J. Bockheim, J.G. Canadell, et al. (2008). Vulnerability of permafrost carbon to climate change: Implications for the global carbon cycle. *BioScience*, 58 (8), 701–714.

Turetsky, M.R., R.K. Wieder, and D.H. Vitt (2002). Boreal peatland C fluxes under varying permafrost regimes. *Soil Biology and Biochemistry*, 34, 907–912.

Turetsky, M.R., R.K. Wieder, D.H. Vitt, R.J. Evans, and K.D. Scott (2007). The disappearance of relict permafrost in boreal north America: Effects on peatland carbon storage and fluxes. *Global Change Biology*, 13, 1922–1934.

Walter, K.M., S.A. Zimov, J.P. Chanton, D. Verbyla, and F.S. Chapin III (2006). Methane bubbling from Siberian thaw lakes as a positive feedback to climate warming. *Nature*, 443, 71–75.

CHAPTER 9: THE CRYOSPHERE AND CLIMATE CHANGE

Further Reading

Review papers by Cooper et al. (1994) and Zachos et al. (2001) provide insight into the Cenozoic evolution of climate and the history of the Antarctic ice sheet. The text by Imbrie and Imbrie (1979) offers a splendid account of glacial–interglacial cycles, unusual in its capacity to blend scientific detail with readability. Hemming's (2004) review article describes the paleoceanographic record of Heinrich events in great detail (see chapter 8), and the American Geophysical Union monograph edited by Clark et al. (1999) contains a good collection of research on millennial-scale climate and cryospheric variability during glacial periods.

There is a large body of literature documenting recent cryospheric change. The IPCC reports contain comprehensive summary information on cryospheric trends and rates of change (Lemke et al., 2007, listed in chapter 1). The Arctic Climate Impact Assessment (ACIA, 2004) and the UNEP report on the state of the cryosphere also provide detailed reviews, and AMAP (2009) provides an update for the Greenland Ice Sheet (see chapter 1). Arctic Council assessments of other aspects of the Arctic cryosphere are forthcoming as part of the Snow, Water, Ice and Permafrost in the Arctic (SWIPA) effort.

Arctic Climate Impact Assessment (2004). *Impacts of a Warming Arctic: Arctic Climate Impact Assessment.* Cambridge University Press, Cambridge, UK.

Clark, P.U., R.S. Webb, and L.D. Keigwin, eds. (1999). *Mechanisms of Global Change at Millennial Time Scales.* Geophysical Monograph. American Geophysical Union, Washington, DC, 394 pp.

Cooper, A.K., P.F. Barker, P.-N. Webb, and G. Brancolini (1994). The Antarctic continental margin—the Cenozoic record of glaciation, paleoenvironments and sea-level change. *Terra Antarctica*, 1/2, 236–480.

Imbrie, J., and K.P. Imbrie (1979). *Ice Ages: Solving the Mystery.* Harvard University Press, Cambridge, MA, 224 pp.

Zachos, J., M. Pagani, L. Sloan, E. Thomas, and K. Billups (2001). Trends, rhythms, and aberrations in global climate 65 Ma to present. *Science*, 292 (5517), 686–693.

References Cited

The quotation from Wallace Stegner is from:

Stegner, W.E. (1987). *Crossing to Safety.* Penguin Books, New York, 341 pp.

My definitions of geologic time come from Gradstein et al. (2005). Snowball Earth events are nicely discussed by Hoffman and Schrag (2002). The evolution of Cenozoic and Pliocene–Pleistocene climate are examined in Zachos et al. (2001) and Lisiecki and Raymo (2005) (figure 9.1). Pollard and DeConto (2009) present models of West Antarctic ice sheet sensitivity to orbital variations. Bond et al. (1997) and the text edited by Clark et al. (1999) provide detailed insight into millennial-scale climate variability during the Quaternary glaciations.

Bond, G., W. Showers, M. Cheseby, et al. (1997). A pervasive millennial-scale cycle in North Atlantic Holocene and glacial climates. *Science*, 278, 1257–1266.

Gradstein, F.M., J.G. Ogg, and A.G. Smith, eds. (2005). *A Geologic Time Scale 2004.* Cambridge University Press, Cambridge, UK, 610 pp.

Hoffman, P.F., and D.P. Schrag (2002). The snowball Earth hypothesis: Testing the limits of global change. *Terra Nova*, 14, 129–155.

Lisiecki, L.E., and M.E. Raymo (2005). A Pliocene-Pleistocene stack of 57 globally distributed benthic $\delta^{18}O$ records. *Paleoceanography*, 20, PA1003.

Pollard, D., and R.M. DeConto (2009). Modelling West Antarctic ice sheet growth and collapse through the past five million years. *Nature* 458, 329–332.

Zachos, J.C., M. Pagani, L. Sloan, E. Thomas, and K. Billups. (2008). *Cenozoic Global Deep-Sea Stable Isotope Data.* IGBP PAGES/World Data Center for Paleoclimatology Data Contribution Series # 2008-098. NOAA/NCDC Paleoclimatology Program, Boulder, CO.

In addition to the ACIA, IPCC, and SWIPA reviews, trends of cryospheric response to climate change are drawn from Kwok and Rothrock (2009) (see chapter 5), Dyurgerov (2002, 2005) (see chapter 6), and the following sources:

Arendt, A.A., K.A. Echelmeyer, W.D. Harrison, C.S. Lingle, and V.B. Valentine (2002). Rapid wastage of Alaska glaciers and their contribution to rising sea level. *Science*, 297, 382–386.

CAPE Last Interglacial Project Members (2006). Last interglacial Arctic warmth confirms polar amplification of climate change. *Quaternary Science Reviews*, 25, 1383–1400.

Cuffey, K.M., and S.J. Marshall (2000). Sea level rise from Greenland Ice Sheet retreat in the last interglacial period. *Nature*, 404, 591–594.

Kaser, G., J.G. Cogley, M.B. Dyurgerov, M.F. Meier, and A. Ohmura (2006). Mass balance of glaciers and ice caps: Consensus estimates for 1961-2004. *Geophysical Research Letters* 33, L19501.

Magnuson, J.J., D.M. Robertson, R.H. Wynne, et al. (2000). Ice cover phenologies of lakes and rivers in the Northern Hemisphere and climate warming. *Science*, 289 (5485), 1743–1746.

Meier, W.N., J.C. Stroeve, and F. Fetterer (2006). Whither Arctic sea ice? A clear signal of decline regionally, seasonally and extending beyond the satellite record. *Annals of Glaciology*, 46, 428–434.

Osterkamp, T.E., and V.E. Romanovsky (1999). Evidence for warming and thawing of discontinuous permafrost in Alaska. *Permafrost and Periglacial Processes*, 10, 17–37.

Prowse, T.D., and B.R. Bonsal (2004). Historical trends in river-ice break-up: A review. *Nordic Hydrology* 35 (4), 281–293.

Rignot, E. (2006). Changes in ice dynamics and mass balance of the Antarctic ice sheet. *Philosophical Transactions of the Royal Society, Series A*, 364, 1637–1655.

Rignot, E., and R. Thomas (2002). Mass balance of polar ice sheets. *Science*, 297 (5586), 1502–1506.

Solomina, O., W. Haeberli, C. Kull, and G. Wiles (2008). Historical and Holocene glacier-climate variations: General concepts and overview. *Global and Planetary Change* 60, 1–9.

Velicogna, I. (2009). Increasing rates of ice mass loss from the Greenland and Antarctic ice sheets revealed by GRACE. *Geophysical Research Letters*, 36 (L19503), doi:10.1029/2009GL040222.

For the sea level rise discussion, estimates of the number of countries and global population centers bordering on the oceans are from the CIA Factbook (https://www.cia.gov/library/publications/the-world-factbook/geos/ho.html, accessed January 2011). This Web resource is updated weekly.

Index

Ingram Content Group UK Ltd.
Milton Keynes UK
UKHW022316230623
423960UK00015B/500